WITHDRAWN
UTSA LIBRARIES

OCEANOGRAPHY AND MARINE BIOLOGY

*Dedicated
to my Mother and the
memory of my
Father*

H. BARNES
B.Sc., B.A., Ph.D., D.Sc., F.R.S.E.
The Marine Station, Millport, Scotland

OCEANOGRAPHY
AND MARINE BIOLOGY

A BOOK OF TECHNIQUES

Ruskin House
GEORGE ALLEN & UNWIN LTD
MUSEUM STREET LONDON

FIRST PUBLISHED 1959
SECOND IMPRESSION 1960
THIRD IMPRESSION 1968

This book is copyright under the Berne Convention. Apart from any fair dealing for the purposes of private study, research, criticism or review, as permitted under the Copyright Act, 1956, no portion may be reproduced by any process without written permission. Enquiries should be addressed to the publisher

© *George Allen & Unwin Ltd., 1959*

SBN 04 551011 3

*Printed in Great Britain
by Photolithography
by John Dickens & Co. Ltd.
Northampton*

PREFACE

THE reading public has shown a great interest during the past few years in the exploration of the sea. In many of the books on this subject the reader finds factual statements, but is left in ignorance as to how they were established. This book is primarily addressed to those who wish to learn something of how our knowledge of the sea is obtained. In it will be found an account of the apparatus used in many branches of oceanography. Although we are not really concerned with results, lest the description of apparatus becomes too tedious, some reference to them—particularly when they are of general interest or when they are rather new—has been made. The descriptions are accompanied by diagrams, either specially prepared or taken from original papers.

There is no recent book of similar scope and original papers have in all cases been consulted; in addition to being of interest to the general reader it may be useful in gathering together for workers in the marine sciences an account of their techniques. To the beginner it will perhaps suggest the instrument most likely to suit his immediate needs; to the student it may serve as an introduction to the subject.

There are two obvious omissions. First, fish populations are sampled by the methods of the fishing industry, and the subject is so vast that any treatment in a volume such as this would be superficial in the extreme; no attempt has therefore been made to give any account of fishing gear, which has been adequately dealt with elsewhere. Secondly, no mention has been made of manned deep diving equipment such as the bathysphere or bathyscaphe. Both are of considerable interest and involve much engineering skill but, as yet, neither has been extensively used as a research tool, although both have distinct possibilities. In any event, they have been described in most readable books—Beebe's *Half Mile Down* and a recent account of the French bathyscaphe work in Houot and Willm's *Two Thousand Fathoms Down*.

I wish to record my thanks to a number of friends, in particular to Mr. B. B. Parrish, who not only read the whole of the manuscript and made many valuable suggestions, but who has constantly encouraged me when enthusiasm or interest waned. Mr. R. E Craig, Mr. K. M. Rae and Mr. R. S. Glover have made many useful comments on some of the sections and Dr. J. N. Carruthers has been most helpful with references. I have had the opportunity of discussions during the writing with my colleague Mr. J. Goodley, and the many friendly arguments have materially served to clarify the descriptions of various pieces of equipment. Others have generously provided unpublished photographs:—

Dr. W. S. von Arx, Woods Hole Oceanographic Institution;

Mr. R. E. Craig, Scottish Home Department, Marine Laboratory Aberdeen;

Dr. G. J. Eicher (jr.), U.S. Fish and Wildlife Service;

Professor K. O. Emery, University of Southern California, Los Angeles;

Dr. N. Ingram Hendey, The Central Metallurgical Laboratory, Emsworth;

Dr. Bruce C. Heezen, Lamont Geological Observatory, Columbia University, New York;

Dr. Martin W. Johnson, Scripps Institution of Oceanography, La Jolla;

Professor B. Kullenberg, Oceanografiska Institutet, Göteborg;

Dr. A. D. McIntyre, Scottish Home Department, Marine Laboratory, Aberdeen;

Mr. K. M. Rae, Scottish Marine Biological Association, Oceanographic Laboratory, Edinburgh.

These are acknowledged in the appropriate place.

I have also to thank my wife, not only for making many drawings, but also for her constant assistance in every way.

CONTENTS

PREFACE   *page* 5

Introduction   13
1. Sampling the Living Organisms   15
2. The Use of Sound Waves   72
3. Some Properties of the Water Itself   106
4. Photography and Television   159

EPILOGUE   202

REFERENCES   204

INDEX   214

ILLUSTRATIONS

1. Water-bottle for collecting samples for bacteriological work *page* 17
2. A marine flagellate—*Chrysochromalina minor* 19
3. Diatoms from the Antarctic 20
4. Hensen net in action 24
5. Large stramin net in action 25
6. Clarke-Bumpus plankton sampler 27
7. Clarke-Bumpus plankton sampler: suspension and closing mechanisms 28, 29
8. Plankton Indicators 30
9. Small high-speed Plankton Indicators 31
10. Stempel (Suction) pipette 32
11. Continuous Plankton Recorders 34
12. Hardy Plankton Recorder: diagram and section 36
13. The modern Continuous Plankton Recorder 37
14. Continuous Plankton Recorder: loading the winding mechanism 38
15. Penetration of *Candacia armata* into the North Sea 43
16. Occurrence of decapod larvae—Southern North Sea 45
17. Beam trawl 48
18. Petersen-type grab 50
19. Holme mud-sampler 51
20. Holme mud-sampler: explanatory drawing 52
21. Smith-McIntyre mud-sampler 53
22. Smith-McIntyre mud-sampler: in use 54
23. Smith-McIntyre mud-sampler: explanatory drawing 55
24. Holme suction grab 56
25. Coring device for short cores 58
26. Under-way bottom sampler 60
27. Under-way bottom sampler: in use 61
28. Emery-Dietz gravity corer 63
29. Emery-Dietz gravity corer: being brought inboard 64
30. Washing a mud sample 65
31. Stetson free-falling corer: explanatory drawing 66
32. Kullenberg piston corer: explanatory drawing 68
33. Kullenberg piston corer: ready for lowering 69
34. Kullenberg piston corer: along side 70
35. Diagram of core from the Atlantic Ocean 71
36. The principle of echo sounding 73
37. Diagram of Kelvin and Hughes-type echo sounder 76
38. Kelvin and Hughes Type M.S. 26 Survey Echo Sounder 77
39. A magneto-striction oscillator 77

40. Outboard unit for echo sounding page 79
41. The Marconi echo sounder 81
42. Cathode ray tube 82
43. Echo-graphs and cathode ray tube records 84
44. Echo-trace of the wreck of the *Lusitania* 85
45. Tracing of a channel during dredging 85
46. Echo traverse in the Antarctic 86
47. Herring shoal as recorded on the 'Seagraph' sounder 89
48. Echo-trace of a series of isolated pilchard shoals 89
49. Effect of light on a shoal of herring: echo-trace 92
50. Effect of light on fish: echo-trace 92
51. Frequency character of croaker noise 97
52. Frequency character of snapping shrimp noise 98
53. Snapping claw of shrimp *Crangon californiensis* 99
54. Change in noise level over a series of traverses 103
55. Change in noise level with distance from shore 104
56. Lumby surface sampler 109
57. Temperature in Western Atlantic Ocean 111
58. Insulated water-bottle 112
59. Insulated water-bottle: closing mechanism 114
60. Reversing thermometers 115
61. Knudsen reversing water-bottle: in action 117
62. Knudsen reversing water-bottle: explanatory drawing 118
63. Mosby thermo-sound 120
64. Spilhaus bathythermograph and sea sampler 121
65. Spilhaus bathythermograph and sea sampler: section 122
66. Spilhaus bathythermograph and sea sampler: mechanism of temperature/depth trace 123
67. Spilhaus bathythermograph and sea sampler: temperature unit and stylus 124
68. Spilhaus bathythermograph and sea sampler: holder for smoked slide, and calibrated scale 124
69. Sea sampler: sample containers 125
70. Sea sampler: triggering mechanism 125
71. Sea sampler: diagram of triggering mechanism 126
72. Sea sampler: closing mechanisms of sample containers 127
73. Drift bottle 130
74. Bottom trailing and double drift bottles 131
75. Routes taken by drift bottles in southern North Sea 133
76. Ekman current meter: explanatory drawing 136
77. Ekman current meter: in use 137
78. von Arx current meters 141
79. Measuring currents with von Arx current meter 142
80. Jacobsen current meters: in use 146
81. Jacobsen current meter: explanatory drawing 147

ILLUSTRATIONS

82. Carruthers drift indicator *page* 149
83. Carruthers drift indicator: explanatory drawing 150
84. Carruthers vertical log 151
85. Geomagnetic electrokinetograph: in use 154
86. Units of geomagnetic electrokinetograph 155
87. Recordings from a geomagnetic electrokinetograph 156
88. Velocity section in north Atlantic Ocean 158
89. Aerial photograph near La Jolla, California 161
90. Beach profiles determined by aerial photography 165
91. Aircraft of type used for visual surveys 168
92. Aerial photograph of red salmon in Brooks River, Alaska 169
93. Underwater photograph of the insular shelf east of Catalina Island 170
94. Underwater photograph of outcrop of eocene 171
95. Underwater photographs of sea bottom at various depths 172
96. Underwater camera for moderate depths 176
97. Underwater photograph of shell-covered bottom in Loch Creran 177
98. USNEL deep sea camera and lighting unit: diagrammatic 180
99. USNEL deep sea camera: in use 181
100. Benthograph 182
101. The scanning process 188
102. C.P.S. Emitron tube and the Millport underwater television camera 190
103. The Millport underwater television camera: underneath view 191
104. The Millport underwater television equipment: inboard units 192
105. The Millport underwater television equipment: on deck of M.V. *Calanus* 195
106. The Millport underwater television equipment: on board M.V. *Calanus* 196
107. Underwater television picture: inshore ground, Firth of Clyde 198
108. Underwater television picture: muddy bottom, Firth of Clyde 198
109. Underwater television picture: gravelly bottom, English Channel 199
110. Underwater television picture: horse-mackerel, English Channel 199

ACKNOWLEDGEMENTS

I WISH to thank the following publishers, journals and authorities for permission to reproduce, in part or entirety, plates or figures from their publications:

Association internationale d'Océanographie physique.
The Biological Bulletin, Woods Hole.
The Bulletin of the Geological Society of America.
The Council of the Marine Biological Association of the United Kingdom.
Deutsches Hydrographische Zeitschrift.
The Editors of the Bulletin of Marine Ecology.
Her Britannic Majesty's Stationery Office.
The International Council for the Exploration of the Sea, Copenhagen.
Messrs. Kelvin and Hughes.
The Journal of the Accoustical Society of America.
The Journal of Marine Research.
The Journal of Wildlife Management.
The Marconi International Marine Communication Co. Ltd.
The McGraw-Hill Book Company, Inc.
The Ministry of Agriculture, Fisheries and Food, Lowestoft.
The National Institute of Oceanography.
Papers in Physical Oceanography and Meteorology, Woods Hole.
The Royal Society, London.
The Scientific Monthly.
Svenska Hydrografisk-biologiska Kommissionens Skrifter.
The University of Chicago Press.

INTRODUCTION

THE oceans cover some seven-tenths of the earth's surface and, if excuse be needed, surely this alone is adequate to justify their study. Oceanography—the science of the seas—is a cosmopolitan activity, for in its broadest sense the term is taken to include all studies of the oceans, their boundaries and their content of living organisms. In reality, the subject cannot claim to be a science in its own right since it has no distinctive discipline. When concerned with the sea as a fluid in motion subject to internal and external forces, we are dealing with a branch of physics, or, in its more theoretical aspects, of applied mathematics. When dealing with the air-sea boundary and reactions between the two milieux, we are concerned with a facet of meteorology. Those who study the sediments and the rocks beneath the seas would regard themselves as geologists, and those whose primary concern is with the chemical composition of the water, the sediments or the living material, as chemists. The living content of the oceans is the province of marine biology, a vast subject in itself, yet still only a branch of zoology and botany.

It is possible and indeed common to make a valuable contribution to any one aspect of oceanography without reference to the others. The physical oceanographer has problems enough to interest him and to tax his skill and ingenuity without referring them to the behaviour of living organisms within the fluid medium. The biologist may study the structure and physiology of the organisms with little knowledge of, or reference to, physical oceanography. However, in ecological problems—problems of the relation of organisms to one another and to their inanimate environment—many aspects of the subject meet, and it is usually unwise and may even be disastrous not to have at least a working knowledge of contemporary activities in many fields. How else can one avoid postulating movements of animals by currents long since shown to be non-existent by physical oceanographers? How can the evolutionist know of the conditions under which animals evolved if he is unaware of modern work on marine sediments?

We have stated that oceanography has no discipline of its own; it has, however, problems of technique peculiar to itself. These are related to the fact that the observations must largely be made and the material, hidden from direct view, largely collected at a distance from the observer. Moreover, all this must usually be done from a ship, often under very adverse conditions; such a moving platform can hardly be considered the simplest place from which to manipulate scientific instruments. As a consequence, the instruments of oceanography must have some mechanism by which the result is either recorded *in situ* and retained for reading when withdrawn from the sea, or means must be provided whereby the information received by the instrument is transmitted back to the vessel. A further consequence of the working conditions is that a robust instrument is greatly to be preferred and this is particularly true of apparatus for collecting much biological material.

A great deal of ingenuity has gone into devising such instruments and these, the tools without which advances cannot be made, will be considered in the following pages.

1

SAMPLING THE LIVING ORGANISMS

THE free-living or so-called pelagic organisms of the sea, that is, those neither attached to nor living within the bottom, encompass a vast assembly of animals and plants, and range from unicellular microscopic bacteria to those highly specialized giant aquatic mammals, the whales. In spite of this, the principles used in sampling them are relatively few; the majority are caught by nets or traps, some by hook and line, others by harpoon; the actual forms that the techniques take are, however, many and varied. For purely qualitative work, for example in studies of form and anatomy, it matters little how the animals are caught. The only criterion of success is that they are caught in adequate numbers and in a sufficiently undamaged condition for the particular problem in hand. It is when we wish to make quantitative investigations that difficulties arise in sampling these free-living forms.

In this section we shall consider these problems in relation to the smaller free-living forms, those which make up much of the so-called plankton. It is difficult to give a rigid definition of this term; in general, it may be considered to cover those small animals and plants (often present in enormous numbers) with restricted powers of movement. They are, therefore, carried about largely under the influence of currents.

The principle commonly involved in collecting these organisms is that of the tea-strainer—with this difference; most frequently the strainer, a net, is pulled through the water and not the water poured through the net. There are, however, some unicellular organisms so small that they pass through the finest nets. These, the bacteria, the so-called nannoplankton, and the smaller diatoms, are often present in enormous numbers and are of fundamental importance in the economy of the sea. They are best counted in water samples and the technique involved will be discussed before considering net-caught plankton.

The bacteria are of various types, but are commonly rod-like, some 2-3 μ long and 0·5 μ thick; the μ-flagellates, constituting the smaller nannoplankton organisms, have somewhat arbitrary size limits—say 1-10 μ—and are elongated, flagellated organisms; the diatoms, which take many curious shapes, are small unicellular plants with a siliceous skeleton.

THE BACTERIA AND NANNOPLANKTON

For enumeration of bacterial content, a sample is usually taken in some form of water-bottle which is let down on a wire run out over a measuring wheel. The depth from which it has been taken and the volume of the sample are, therefore, known. Dr ZoBell has, however, suggested that the ordinary metal water-bottles used for salinity samples (see pp. 110-119) are not entirely satisfactory for bacteriological work; probably as a result of the slight solubility of metals in sea water, there often appears to be a bactericidal (bacteria killing) or bacteriostatic (bacteria growth-preventing) effect in samples from such standard waterbottles. Further, they cannot be maintained sterile before the sample is taken, so that, as a result of the ubiquity of bacteria, there is great risk of contamination. For accurate work, he recommends the use of a glass, rubber, or plastic bottle fitted with a glass tube: the whole apparatus can be sterilized (by the standard procedure of autoclaving) and the capillary end of the glass tube then sealed. Either one or a series of such bottles, mounted as shown in Figure 1a, is attached to a hydrographic wire (in the same way as ordinary water-bottles). A

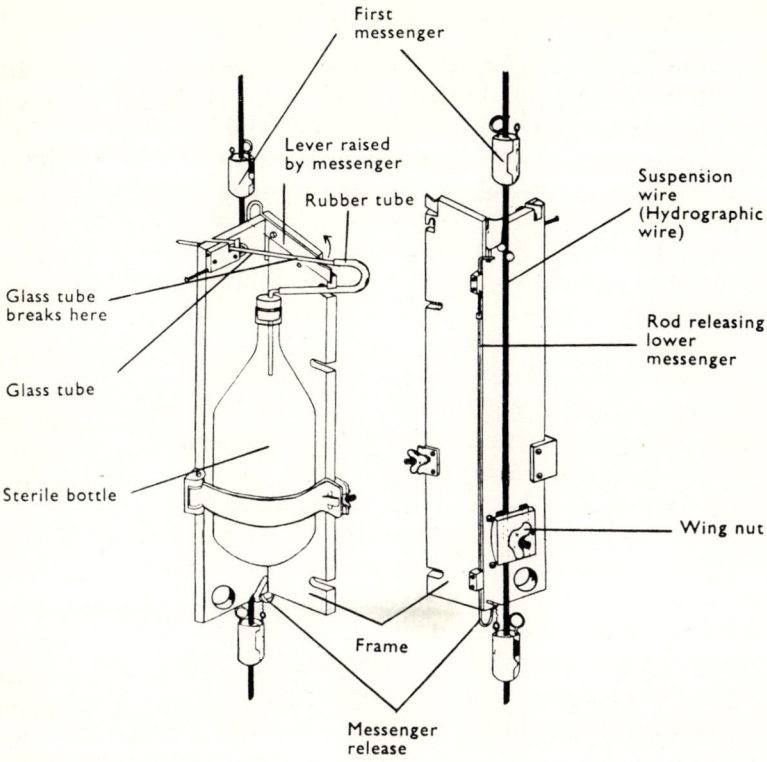

Figure 1
a) Water-bottle for collecting samples for bacteriological work: front and back views. The messenger is shown about to strike the lever which will break the glass tube and also move the rod releasing a second messenger

(*C. E. ZoBell*, '*Journal of Marine Research*', 1941; *Redrawn from original*)

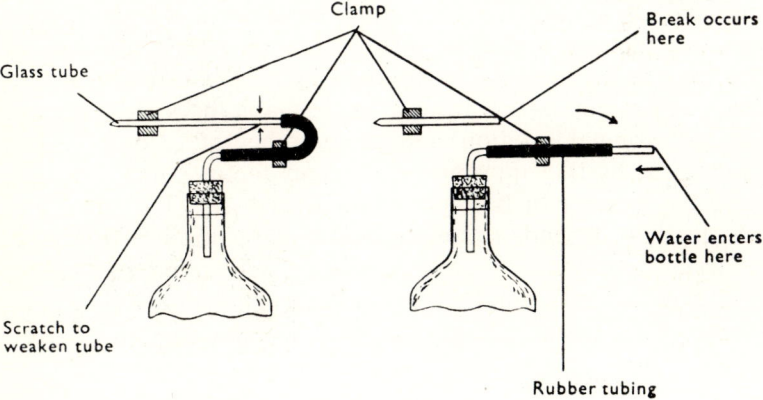

b) Breaking the glass tube in the ZoBell bottle

'messenger', that is, a brass weight, is sent down the wire and strikes the lever; this puts a strain on the glass tube, which breaks where it has been weakened by a scratch mark. The free end, because of tension in the rubber connection, flies to one side and water is sucked into the bottle as a consequence of the partial vacuum resulting from sealing it immediately after autoclaving (Figure 1b). As with water bottles in hydrographic work, each messenger can be made to release another which actuates the next bottle so that a whole series of samples may be taken at different depths on a single wire.

Once a sample of sea water is obtained, how is the number of bacteria present determined? There are several methods—each with advantages and disadvantages and, for detailed work, it may be necessary to apply each one successively to the same sample. The bacteria may be counted directly; a small portion of the sample is transferred to a chamber whose volume is known and the bacteria in a given fraction of this volume counted under the high power of a microscope. (This is the procedure used for counting blood cells, and the counting chamber, ruled into squares, is termed a haemocytometer.) Multiplying by an appropriate factor gives the bacterial content per unit volume. This method, while giving a reliable estimate of the numbers present, does not indicate the types of bacteria; furthermore, all those counted may not be living. As an alternative, the so-called plating method may be used—a standard technique in bacteriological work. In this method a sample of known volume (that is, a known fraction of the original) is transferred to culture medium and spread on a gelatine plate in a shallow dish. After some time, colonies of bacteria grow and may be counted by eye: it is assumed that each colony has arisen from a single viable bacterium present in the original solution and the number of colonies, when multiplied by the appropriate dilution factor, will, therefore, give the number of bacteria originally present. The success of the method depends on providing, in the culture medium, the nutrients required for growth of the bacteria present in any particular sample under investigation. The nutritive requirements of different bacteria vary widely, so that it is not possible

SAMPLING THE LIVING ORGANISMS

to get a single nutrient medium adequate for all types. However, by using a series of different nutrient media in successive tests something is also learnt of the types, as well as numbers, of bacteria present.

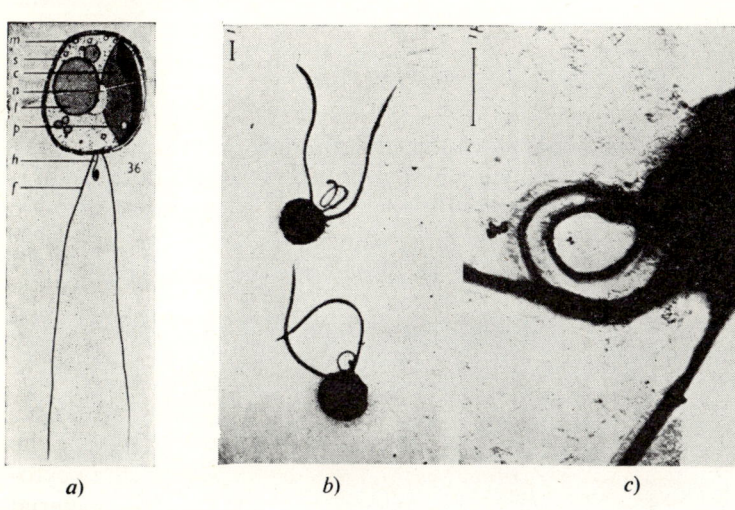

a)   b)   c)

Figure 2
A marine flagellate—*Chrysochromalina minor*

a) A diagram of the single celled organism,
        c—chromatophore m—muciferous body
        f—flagellum n—nucleus
        h—haptonema p—pyrenoid-like body
        l—leucosin vesicle s—scales

b) and c) Electron micrographs at different magnification showing flagella and the coiled haptonema. The latter is possibly connected with feeding activities

(*From M. Parke, I. Manton and B. Clarke, 'Journal of the Marine Biological Association, U.K.', 1955*)

The μ-flagellates or nannoplankton (Figure 2), like the bacteria, have no skeleton; they are very delicate and fixatives do not always leave them in a recognizable state. They are, therefore, best counted alive. In samples from the open sea, as distinct from experimental cultures, they must usually be concentrated before counting. Some workers do this by centrifuging the sample, others prefer to pass a known volume of sea water through a filter with extremely fine pores; just before the filter becomes dry, the concentrate is transferred to a counting chamber. In order to recognize and name the species counting is best done by daylight, again under high-power magnification.

THE DIATOMS

Diatoms (as well as some microscopic protozoa such as Peridinians) are likewise best sampled by water-bottle. However, these organisms, with their strong skeleton of silica or calcium carbonate, are more robust and may be preserved before counting (Figure 3). When the numbers are high, a

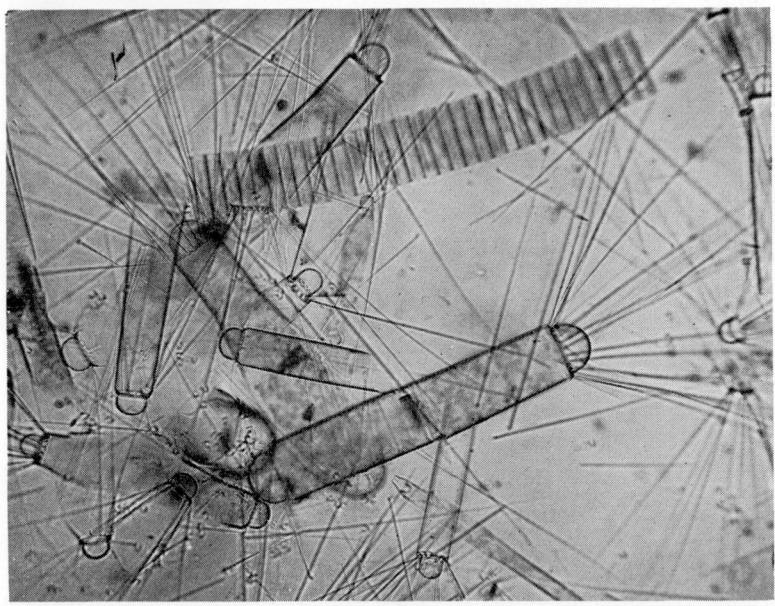

Figure 3
Diatoms from the Antarctic. The dominant species is *Corethron criophilum*, rather like a spinous cocoon. This plankton is the basis of food for the whales
(*Courtesy of Dr. N. Ingram Hendey*)

SAMPLING THE LIVING ORGANISMS

known volume from a water sample is transferred directly to a haemocytometer cell and the diatoms counted in the same way as bacteria. When the numbers are low, a larger volume of water may be first centrifuged in order to concentrate the organisms, and this may be facilitated by the formation of a slight precipitate in the water; this helps the diatoms to sediment under the centrifugal force. An alternative method is to kill the diatoms in a known volume contained in, or transferred to, a tube with a thin microscope cover slip as its base. This is allowed to stand on the stage of an inverted microscope (which has its stage above the objective) and, after the diatoms have settled, they may be counted without disturbance under high power from below.

Even over small areas of the sea the numbers of these unicellular organisms vary widely and counts of a single water sample, whilst valid for that particular sample, may give only a poor representation of the population over the whole area being studied. In consequence, the only accurate way to get a reliable estimate of the numbers is to take a very large number of samples over the area. Then, not only may an average abundance be calculated but—and this is important—the 'spread' of the values will also be known. When water samples are taken, this involves much work.

Fine nets are also used to collect diatoms and, particularly if one is only concerned with the bigger species and with comparative rather than absolute numbers, the results can be useful. To obviate the very considerable task in counting these organisms, other methods have been evolved.

The diatoms, like land plants, contain pigments, principally chlorophyll, and Dr Harvey of the Plymouth Laboratory utilized this fact to develop a method of estimation. The diatoms are collected by hauling a fine net through the water and, after being separated by filtration, the pigments are extracted with an organic solvent. The coloured extracts are then matched against standard tints. The results are usually expressed in 'pigment units'. From preliminary experiments, in which a comparison is made between counts and colour intensities, the latter may be converted into the approximate number of diatoms. This method tells us nothing about the composition of the diatom flora—what species are present and how many there are of each—and it is subject to inaccuracies due to variation in the intensity and kinds of pigments present. Nevertheless, as a comparative method which is simple and rapid, it is important and has been further developed and extensively used in certain types of investigation.

ZOOPLANKTON

And now we come to the zooplankton, that is, the animal plankton. The larger plankton organisms are usually caught by nets—the problems associated with the quantitative use of which were first tackled in the early part of this century by the German school of plankton workers, the physiologist Victor Hensen being particularly prominent.

The first problem is to determine how much water has been filtered in any given tow—from what volume, in fact, the catch was taken. The quantity of water passing through the net depends principally upon the mesh size, the area of the filtering surface, the shape of the net, the area of its opening and the speed of towing. Hensen made a long series of careful experiments to determine the effect of some of these variables and calculated a so-called filtration coefficient. This expresses the amount of water which passes through a net in relation to the amount which would pass through the mouth if no net were attached; the latter is, of course, the area of the opening multiplied by the length of tow.

'Correction' of results by a filtration coefficient has been widely used, but there is an important factor affecting the amount of water filtered for which no allowance can be made in this way. This is the effect of the organisms themselves. These, particularly when a fine net is being used and when diatoms are present in large numbers, soon clog the meshes and the filtering capacity steadily decreases during the course of a tow. Since this cannot be allowed for by calculation, modern practice is tending more and more to incorporate a flow meter in the mouth of the net. This registers the number of turns of

SAMPLING THE LIVING ORGANISMS

a propeller, and after calibration (in separate experiments) the meter reading in any haul gives the amount of water which has passed through the net. Even so, the method is not entirely satisfactory, since one never knows exactly at what point clogging begins; further, clogging may affect the catch differentially—that is, the effect may not be identical for different organisms.

PLANKTON NETS AND PUMPS

Large plankton nets usually take a common form (Figures 4 and 5). They consist of the net proper attached to a ring by a canvas band at the upper end and at the lower end by a second canvas band to a collecting vessel. To cut down water pressure and reduce the opening relative to the filtering surface, a conical non-filtering portion may be added in the upper part, and this is fastened to a smaller ring and stiffened by supporting rods. The collecting bucket takes various forms. It is best made with a gauze window so that the water runs out as the plankton is washed down into the bucket. In some types the main part of the bucket may be detached from the net and the collected material run off through a stopcock at the bottom.

In the early work many samples were taken by what are termed vertical hauls (see Figure 4); with the ship stopped, the net is let down on a wire and then slowly drawn up through the water. In this way the water column as a whole is sampled. However, this method suffers from the same disadvantages that we have discussed in connection with the estimation of diatom populations from water-bottle samples; unless a large number of samples are taken, we may only get a poor representation of the plankton content over an area.

Horizontal hauls taken over a considerable distance reduce such 'errors'. They do, however, introduce two new problems. First, the depth at which the net is towed must be known. If the speed of the towing vessel can be accurately controlled, then the depth of the net for a given towing speed may be determined in a series of trial runs, using a depth-recording device attached

OCEANOGRAPHY AND MARINE BIOLOGY

Figure 4
Hensen net in action. This fine net is used for catching small zooplankton organisms. The fine mesh is surrounded by a coarser net for protection. Notice the non-filtering cone at the top above the net and the collecting bucket just clear of the water. The net is being worked by means of a counter-weight

(*By permission of the Ministry of Agriculture, Fisheries and Food, Lowestoft*)

SAMPLING THE LIVING ORGANISMS

Figure 5
Large stramin net coming to the surface. This is made from fairly coarse material and is used to catch the larger zooplankton organisms, since only these are retained by the coarse mesh. Note the trawl floats that keep the head up
(*By permission of the Ministry of Agriculture, Fisheries and Food, Lowestoft*)

to the net; indeed, such a device may be attached every time a haul is made. However, modern practice is tending more to the use of a depth-keeping device, either on the net itself or on the towing wire to which the net or a series of nets is attached. The paravane principle is often employed. The apparatus has a plane (or planes) inclined to the direction of tow; the thrust of the water during towing forces it down and, within wide limits, it maintains itself at a fixed depth for a given amount of warp out independently of towing speed; as this increases, the tendency for it to rise is overcome by the increased downward thrust on its inclined diving planes.

The second additional problem introduced when horizontal hauls are used centres round the opening and closing of the net at the required depth; many devices have been suggested for this purpose. Ideally the net must remain closed during descent, must be capable of being opened at the required depth, and then closed at the end of the tow before hauling in; if possible, there should be no cumbersome mechanism obstructing the opening. Some of the early closing devices, often depending upon water pressure, were not very 'positive' in their action and some, depending on complicated mechanisms, were very liable to go wrong. The closing net of Kofoid seems very satisfactory although it does not appear to have been much used. The net itself was attached to a ring hinged into two halves actuated by springs. Messengers were used to open and close the jaw-like entrance. In some modern instruments the 'butterfly' valve is used and will be described in connection with a modern net.

The Clarke-Bumpus sampler is a modern closing net now widely used (Figures 6 and 7) and incorporates many desirable features for quantitative work. It consists essentially of a short brass tube, 5 in. in diameter and about 6 in. long, to one end of which a net (about 2 ft. long) of the desired mesh is attached by means of a ring with a bayonet lock. The tube is mounted on a metal frame fastened directly to the supporting cable by means of pivots that allow the tube to turn in a vertical plane; when towed, the net always takes up a horizontal position regardless of the cable angle. Towards the rear two metal planes, mounted at an angle to the tube, keep the instrument stable

SAMPLING THE LIVING ORGANISMS

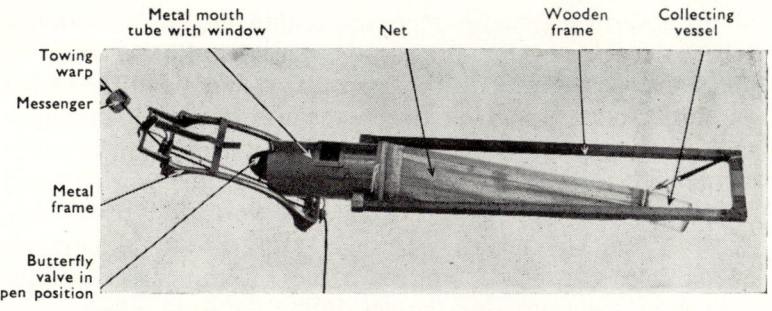

Figure 6

a) The Clarke-Bumpus plankton sampler. The 'lid' to the metal fore-part is in the open position and the messenger is about to strike and close the lid

b) Details of the metal fore-part. The water meter can be seen through the celluloid window. Note method of swinging net from frame

during towing. The frame is attached to the cable by means of a loosely fitting spring pin at the top and at the bottom by a gate lock which closes around the neck of the supporting clamp, thus securing it rigidly to the cable. In this way the frame is free to swivel round the cable so that during towing the opening of the tube is always directed forward (Figure 7a).

A shutter in the form of a circular disc pivoted to rotate

about a vertical plane is mounted within the entrance of the tube: when this is closed, no water can flow into the net. This metal disc is set against springs so that it may be opened from an initial closed position (when the net is sent down) by a messenger, the disc turning through 90°. When the tow has been completed, a second messenger actuates another spring catch and the 'lid' is again closed by turning through a further 90° (Figure 7b). The net may then be hauled without catching animals. A meter is mounted in the tube and registers the volume of water entering the net during the tow.

All the nets so far considered are towed at relatively slow speeds in order to avoid breaking the fine-meshed net material. There is evidence that some of the larger planktonic animals are able to avoid capture by nets towed slowly—either in response to water pressure in front of the net, or by seeing it and taking avoiding action. Such losses may be avoided by using high-speed nets: in these, the net surface is large in relation to the opening so that back pressure is reduced; the net is usually fitted into a torpedo-shaped body, which is towed through the water at high speed. A simple high-speed net is the Hardy

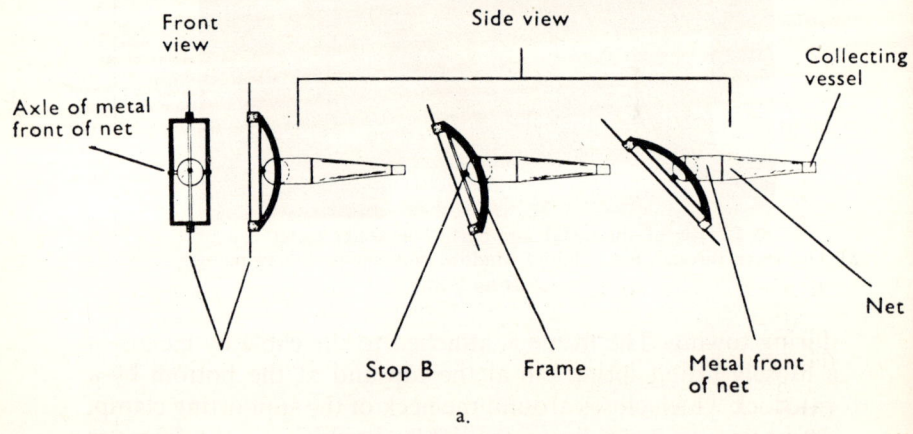

Figure 7

The Clarke-Bumpus sampler

a) The suspension net pivoted on frame, so free to tow horizontal whatever angle of warp

SAMPLING THE LIVING ORGANISMS

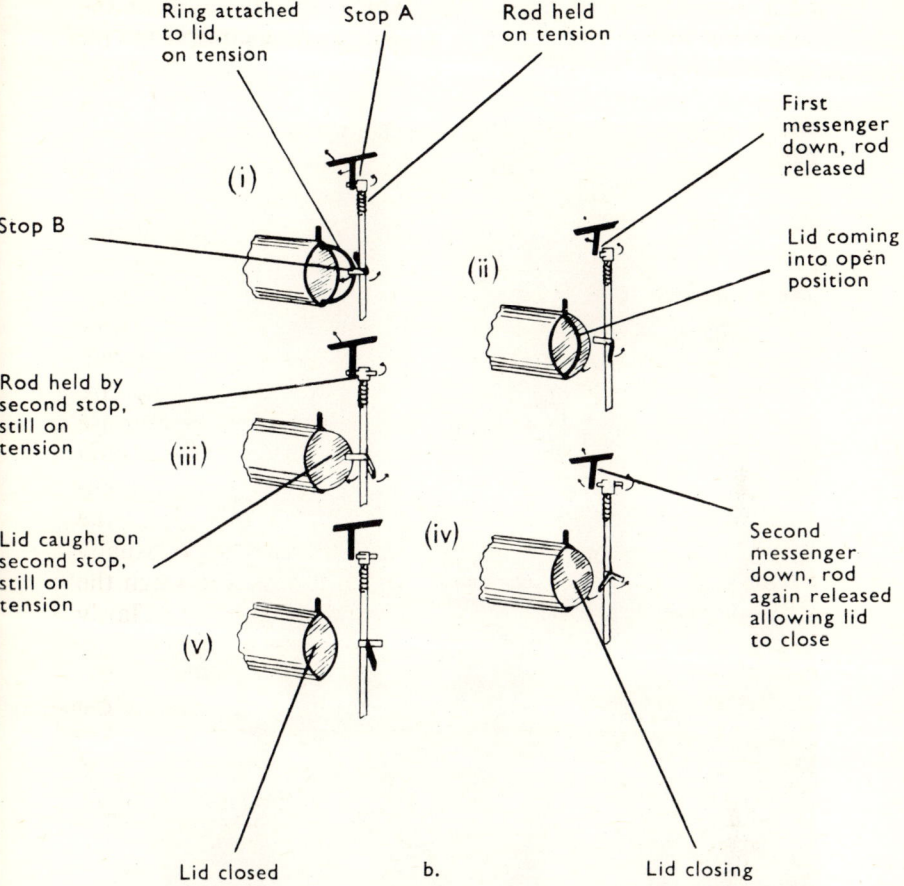

b) The release mechanism:
  (i) Set for going down; butterfly valve closed, stop against semi-circular ring
  (ii) First messenger down; allows rod to rotate releasing ring
  (iii) Open; lid has caught on second stop at 90° to first stop; second stop has caught upper release
  (iv) Second messenger down, again released allowing closure
  (v) Closed

Plankton Indicator (Figures 8 and 9). In some models a disc of netting is mounted on a brass ring and held in the back

OCEANOGRAPHY AND MARINE BIOLOGY

of a torpedo-shaped metal tube, which is itself mounted on a depth-keeping device. In other models a small conical net may be fitted inside the tube.

Figure 8
Plankton Indicators
Three models of various sizes are shown. The gauze disc is being taken out of a larger model. Smaller models on the bench
(*Courtesy of Mr K. M. Rae*)

The net method is the tea-strainer in reverse. It is possible to obviate many of the difficulties associated with tow-nets by

SAMPLING THE LIVING ORGANISMS

reverting to the direct tea-strainer, that is by pouring the water through a net. In this technique water is pumped up from the required depth and passed through a net of appropriate mesh, the net being held in a tank of water to take up the force of the

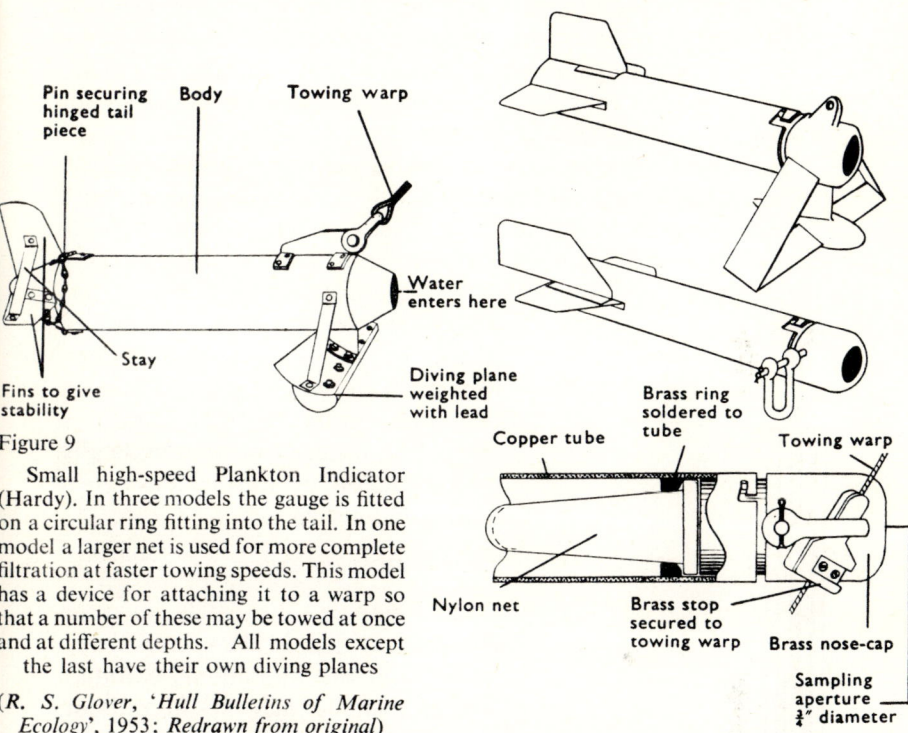

Figure 9

Small high-speed Plankton Indicator (Hardy). In three models the gauge is fitted on a circular ring fitting into the tail. In one model a larger net is used for more complete filtration at faster towing speeds. This model has a device for attaching it to a warp so that a number of these may be towed at once and at different depths. All models except the last have their own diving planes (R. S. Glover, 'Hull Bulletins of Marine Ecology', 1953; Redrawn from original)

delivery. For moderate depths quite small pumps may be used, since the only 'lift' required is that from the sea surface into the boat: apart from this, it is only necessary to overcome frictional resistance in the hose. The volume filtered may be registered by a meter on the pump or determined by filling a tank of known dimensions; both the depth from which the sample is taken and the amount of water filtered are then accurately known. The method is usually used with the ship stationary, but greater areas or depths can be sampled by working with the ship slowly under way and by moving the hose up and down while pumping. A hose is somewhat cumbersome to manage and becomes more so the greater the depths from which samples are taken. However, this difficulty has recently been overcome by dispensing with the hose and using a submersible pump fitted with a water meter and net. There is one further disadvantage of the method; although the smaller animals come through a centrifugal pump undamaged and indeed often alive, the larger ones may be mutilated—on occasions to such an extent that recognition is difficult.

Separate problems arise in the sampling of plankton very close to the bottom, and various nets have been devised specially for this purpose. In one method, a net is mounted on an Agassiz trawl frame (see page 47), which, since it is on runners, may be dragged over the bottom. A more recent bottom net has doors held by springs over the mouth of the net; to the doors are attached long arms and when these are dragged over the bottom they pull the doors open; the latter close automatically by the action of the springs when towing is finished.

Enumerating the catch

Once obtained, the catch must be enumerated, that is, the species determined and the individuals counted. With the smaller nets now in common use the whole catch may be dealt with, but if larger nets are employed then it may have to be sub-sampled and only a fraction counted. This requires that the organisms shall be uniformly distributed throughout the sample and a known fraction taken for counting. There are various ways of such sub-sampling. Hensen devised a special suction pipette. This is really a syringe (Figure 10), the end of the

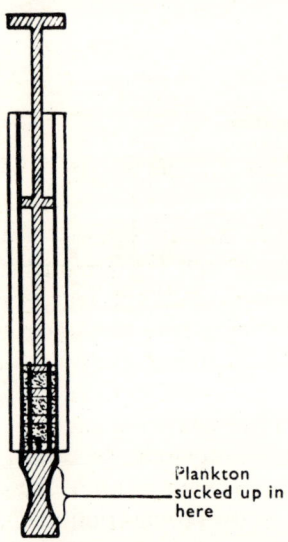

Figure 10
Stempel (Suction) pipette. Used for taking an aliquot from a plankton sample. The sub-sample is contained between the curved part of the plunger and the barrel of the pipette

SAMPLING THE LIVING ORGANISMS

piston being cut away so that a definite volume is enclosed in this cut-away portion when the piston is raised. The plankton in a known volume of water is stirred by means of the syringe and the piston then raised: the fraction taken can be calculated and the animals counted. Another method is to transfer the preserved plankton sample to a 'whirling' vessel; after thoroughly stirring, a set of vanes is dropped quickly into the suspension, dividing it into a number of equal sectors; the contents of one sector are run off at a tap in the base of the container and counted.

THE SYNOPTIC PLANKTON PICTURE—HARDY CONTINUOUS PLANKTON RECORDER

A specially equipped research ship is a necessity for most types of plankton work. However, for the study of some plankton problems a research ship is unable to supply sufficient samples over a wide enough area. Plankton populations are not static; the pattern of their distribution—a veritable mosaic in three dimensions—is continually changing and it is of the greatest importance to know something of these changes. Once we do know something of such changing patterns we can attempt to relate them to other variables, for example, nutrients, temperature, salinity, wind and ocean currents. Further, the populations of the sea are closely interdependent and a study of the changing plankton picture from year to year should, since it gives a picture of changes in the biological environment, yield valuable information in relation to fishery research. Something like the weather charts, now a regular feature of the daily newspapers, is wanted, but showing the plankton distribution in the seas. A map, in which the distribution of a set of variables over an area at a given instant is plotted, is called a synoptic chart. Such a picture giving the distribution over a wide area at any one time is important and, if successive synoptic charts are obtained, we then get a truly dynamic view of plankton populations.

Even if time and money were available, it would be almost impossible for a more or less continuous synoptic programme

Figure 11
Continuous Plankton Recorders
The large recorder in the centre is the original model tried out by Professor A. C. Hardy in the Antarctic. In the background is a smaller recorder taken on Sir Herbert Wilkins' *Nautilus* Expedition. A modern recorder is shown in the foreground
(*Courtesy of Mr K. M. Rae*)

of plankton work to be carried out by research vessels. With the difficulties inherent in obtaining such information in mind, Professor (now Sir) A. C. Hardy designed a plankton-collecting instrument which he called the Continuous Plankton Recorder (Figure 11); this can be worked from merchant ships without difficulty and without any disturbance of their normal routine. Its possibilities were first explored during a *Discovery* Expedition to the Antarctic in 1925-27; these preliminary trials allowed many of the mechanical 'teething' troubles to be eliminated and a smaller, simpler, more robust, and more efficient machine to be designed (Figures 12, 13, 14). On his return, Professor Hardy was able to set up a small team of workers at the University College, Hull, to deal with the samples collected by this new instrument and now the work, with its scope considerably expanded, is continued at Edinburgh as part of the activities of the Scottish Marine Biological Association.

The machine

The principle of the recorder is quite simple. A hollow, more or less torpedo-shaped body is towed through the water which enters at a small opening in the front; the plankton is both filtered off and stored on a silk gauze which continually moves across the incoming stream, the water passing out at the rear of the instrument. The result is a length of silk gauze covered with plankton. The speed of the gauze through the machine and the speed and course of the ship being known, it is then possible to determine the distribution of plankton along the ship's route. When the results from a number of recorder tows, made over different routes at the same or similar times, are entered on a single chart, a close approach to a synoptic picture is obtained.

The machine (Figures 12-14), which consists of an outer shell and an internal mechanism for winding the gauze and storing the reels, is 3 ft. 4 in. long and weighs 156 lb., the weight being largely that of the lead put into the front section to bring the centre of gravity forward: the cross-section is square, so that the internal mechanism may be more conveniently housed. Below and towards the front, strong brackets support an inclined diving plane and this ensures that, with a given length of towing warp, the instrument will tow at a constant depth in spite of considerable changes in the ship's speed (see page 26): variation from 8 to 16 knots has been shown not to affect the towing level of the recorder. As may well be imagined, the strain on the towing cable is considerable, but more important than this is the fatigue of the wire resulting from vibration at the point where the cable is attached to the recorder. The towing cable is, therefore, fastened to the recorder through a

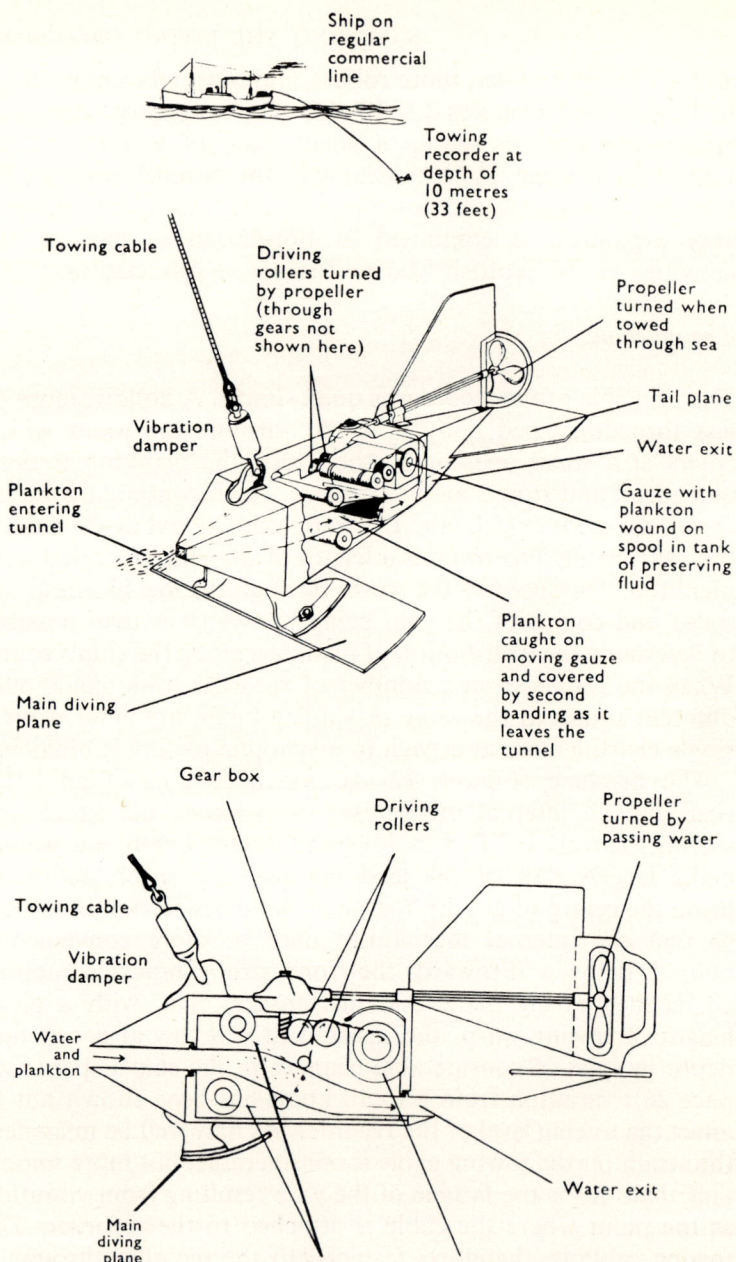

Figure 12

Diagram and simple section of the Hardy Plankton Recorder
(*A. C. Hardy*, '*Hull Bulletins of Marine Ecology*', 1939, 1944; *Redrawn from original*)

SAMPLING THE LIVING ORGANISMS

shock-absorber and this takes the form of a rubber compression block within a steel cylinder. To give the machine stability, small paired horizontal fins as well as a large vertical fin are fitted to the tapered tail section. Inside the vertical fin is a small

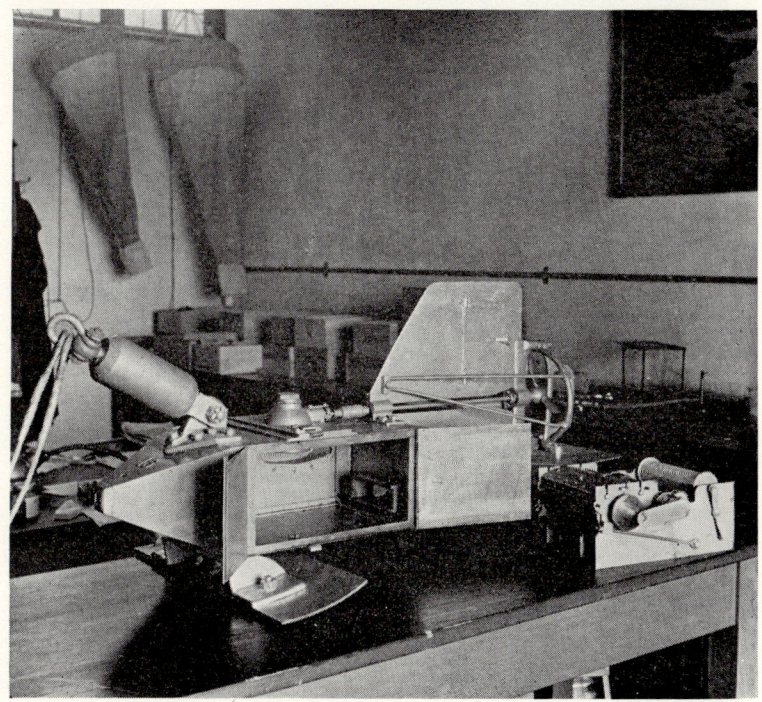

Figure 13
Continuous Plankton Recorder
The modern recorder for use from merchant ships. Note the shock absorber where the towing cable is attached, the diving plane underneath the fore part of the main body, the propeller with its guard, and gear case for spool drive. The removable gauze and winding mechanism is out of the instrument
(*Courtesy of Mr K. M. Rae*)

rudder, set so that the recorder will veer to one side and hence avoid the log line and the disturbed water in the ship's wake.

The motive power that drives the spools drawing the gauze continually across the stream of water is obtained through a propeller mounted on top of the machine. This propeller

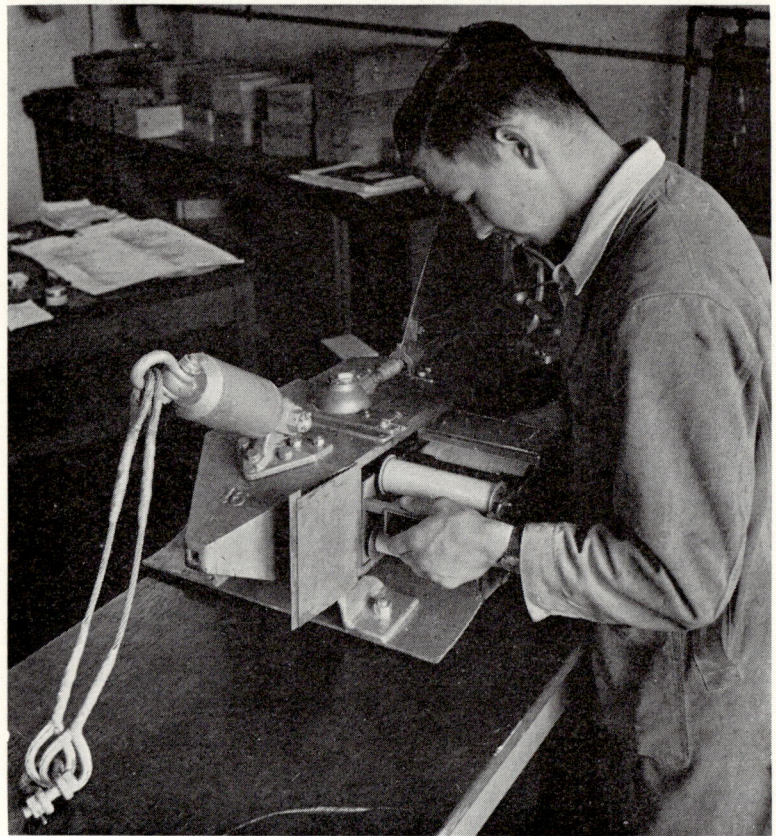

Figure 14
Continuous Plankton Recorder
Inserting the newly loaded winding mechanism into the body of the instrument. Note the square hole in nose for entry of water
(*Courtesy of Mr K. M. Rae*)

rotates rapidly when the instrument is towed. The propeller shaft passes forward through a bearing and connects by means of a flexible joint to a drive-shaft entering a dome-shaped gear-box. By changing the pitch of the propeller blades, the speed of the gauze can be altered; in this way the plankton record may, so to speak, be 'expanded' or 'contracted'. The limit

SAMPLING THE LIVING ORGANISMS

of any such 'expansion' is, of course, set by the amount of gauze which the spools can take and the distance the recorder is to be towed. In the earlier models it was found that the propeller was sometimes jammed by seaweed, and even on one occasion by a decapitated fish! Guard rails were, therefore, fitted round the propeller, and these also serve as handles to facilitate lifting the machine.

Water enters the tunnel at the nose through a $\frac{1}{2}$ in. square hole and passes backwards into the wider portion, which is rectangular in cross-section (Figure 12). The central section of the tunnel is formed by a separate unit containing the gauze spools together with the appropriate gearing and, although when in position it makes a tight joint, this unit slides out at the side of the machine. When the frame is pushed into position a worm from the propeller gear-box engages the gear from the driving spool and a train of gears—via a special conical drive—drives the storage spool.

The collecting gauze, initially all on the lower spool, passes across the tunnel and leaves through a slit in its roof. It is necessary to keep the plankton in the position in which it was collected on the gauze and therefore, as soon as the latter leaves the tunnel, it is overlaid by another gauze coming from a second upper spool. Thereafter the two gauzes, tightly gripped together at the edges only, pass backwards as a 'sandwich' of plankton on to the storage spool in the tank at the rear (Figure 14). This tank contains a preservative, formalin.

After the machines are returned to the laboratory, the spools are taken out, the gauzes unrolled and cut up into 'blocks', as they are called, each equivalent to 10 miles of tow and each, therefore, having on it the plankton from a 10-mile line of water. The preserved organisms may be examined, identified and counted at any time. In order that the results may be accurately plotted on charts, the masters of the towing vessels fill in a standard form; they give log readings when the machine was both launched and hauled as well as any additional check readings on passage. It is usually arranged for the recorders to be shot and hauled at definite points. For example, on the line to Bremen across the North Sea the point for shooting on the outward voyage is the Outer Dowsing Lightship and that for hauling is the Norderray Lightship.

The advisability of running a recorder on any particular voyage is, of course, left to the master of the vessel concerned. They are not run in fog, very severe weather or when the ship is amongst herring nets.

The ships that tow these recorders are provided with a small hand-winch, towing rope, davit and block, and on the first run instructions are given by a member of the plankton team. As the recorder is launched the ship slackens to half speed, the instrument is lifted over the rail and 9 fathoms of cable veered out against the winch brake; the machine will then be towing at the standard depth of 10 metres.

We have stressed that all methods of plankton collection have their limitations—the Continuous Plankton Recorder is no exception. Before looking at these limitations, which might appear seriously to detract from its value, we must, however, stress that at present it is the *only* instrument that will give any approach to serial synoptic plankton pictures—and this with relatively small running costs; in spite of its limitations it is, therefore, a valuable addition to the techniques of plankton research.

Limitation of the technique

It has been stressed that no one net or method is satisfactory for catching *all* types and *all* sizes of organisms at one and the same time and, since a single gauze is used in a recorder, it must suffer from this limitation. Once this is accepted, it is then necessary to decide upon what type of information is needed for the particular investigation in hand; the appropriate gauze mesh is chosen and some information, no matter how desirable for other purposes, must be sacrificed. Perhaps it should be added that this limitation must be clearly kept in mind when interpreting the results. The gauze chosen, 60 meshes to the inch, is a compromise—some small organisms escape, but the large and medium-sized planktonic organisms are retained.

The second disadvantage, namely, clogging of the gauze by the

SAMPLING THE LIVING ORGANISMS

plankton itself with consequent reduction in filtration capacity, is also common to other net methods. However, since the ratio of the area of the opening to that of exposed filtering gauze is high, this will only occur when organisms are taken in very large numbers: in this event it must be remembered that the number counted is to be regarded as minimal. More serious is the fact that there is a differential effect as a result of clogging; this will also affect the catching power in respect to organisms other than those actually responsible for the clogging. For example, suppose the copepod *Calanus* is present in sufficiently large numbers over a 10-mile block to cause clogging and to reduce the catching power by a factor of 10. Then instead of the true value of say 10,000 *Calanus* one would estimate 1,000; this could still be large compared with the numbers of *Calanus* on adjacent blocks. However, another organism present in this region at a true value of say 10 would be reduced to 1, and this could be lower than the catch taken in the absence of clogging in an adjacent patch with the same number present. Further, extremely rare forms which are often of great importance in some problems may be entirely lost as a result of clogging. However, for the purpose and in the way the recorder is used this is relatively unimportant.

A further limitation is that imposed by the single depth of sampling, namely 10 metres, since the distribution obtained is, strictly speaking, only that at this depth. Animals are not uniformly distributed in depth and ideally one would like to have synoptic charts for a number of standard depths. This could, of course, be done by towing simultaneously a number of recorders, one at each of the selected depths. With the present scheme—namely, using merchant vessels—this is hardly a practical possibility. Even if the animals were not distributed uniformly but always remained in the same proportions at the various depths, then the values at any given depth would always be comparable. This is not so; many organisms, particularly the important copepods, make regular diurnal migrations, rising by night and going down by day. When this effect is marked, the results of a tow will be grossly biased and it will be misleading to compare day blocks with night blocks. So far the greater part of this work has been done in the North Sea, and analysis of the results has shown that in this region the effect of vertical migration is not great enough to obscure the general picture. This migration must, however, be borne in mind as the work extends to other regions—particularly where the sampling is over deep water; it soon becomes evident from the results which animals are showing diurnal migration; when regular tows are taken, the night and day blocks are never consistently placed in space so that, even with extensive vertical migration, it becomes possible to eliminate bias as records accumulate. There is another restriction; the necessity to maintain the animals in the position in which they are caught by means of a covering gauze, which together with the collecting gauze is wound on to the spool in the preserving tank, results in the animals being squashed. This tends to make specific recognition of the animals difficult. In spite of this, many are still easily recognizable by their standard zoological characters and for others different features from those normally employed may be used; in some cases new methods of recognition of the somewhat mutilated organisms have been developed.

Application of the technique

What sort of problems may be tackled by recorder surveys and by a study of the synoptic charts so obtained, giving as they do, the pattern of plankton month by month and year by year? Questions such as the following can be answered. Is the seasonal sequence repeated annually? How great are the fluctuations in numbers over any area from month to month and from year to year? Do the various organisms behave differently as regards either the numbers present or their distribution? There is such a wealth of data that it is difficult to choose simple illustrative examples; the first to be given illustrates most clearly the advantages of a synoptic picture in the interpretation of the movement of plankton populations. Figure 15 (taken from a paper by K. M. Rae and C. Rees) shows the positions at which the copepod *Candacia armata* was taken by the recorder during the latter part of 1938, the various lines indicating the south-eastern

SAMPLING THE LIVING ORGANISMS

boundary of its distribution in successive months. As the season progresses it is evident from this work that the animal steadily penetrates farther and farther into the North Sea. Now the species is known to thrive in oceanic water; the steady

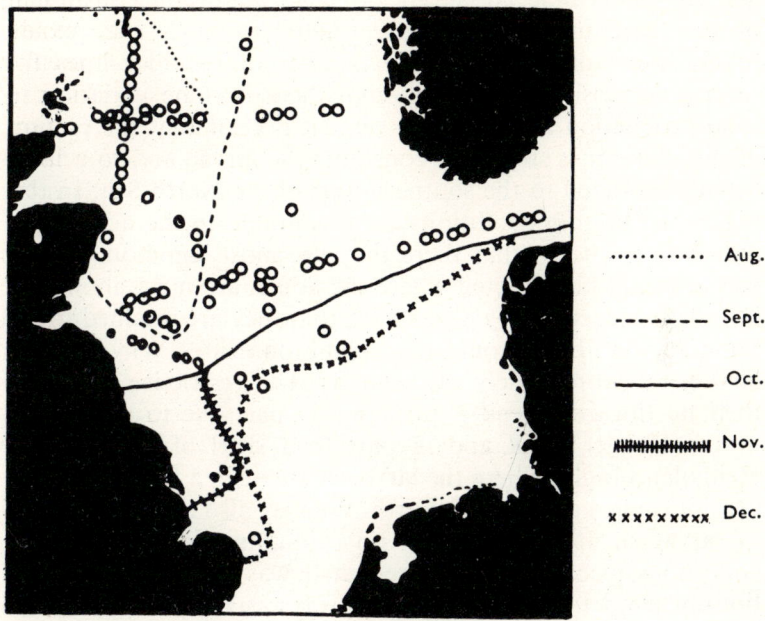

Figure 15
Penetration of *Candacia armata* into the North Sea. Circles indicate the presence of *Candacia armata*, and the various lines the limit of penetration during successive months
(*K. M. Rae and C. B. Rees, 'Hull Bulletins of Marine Ecology', 1947; Modified from original*)

progression southwards of the limit of distribution can therefore be considered to give an indication of the penetration of Atlantic water into the North Sea.

The recorder programme has been concentrated largely on those organisms which, from egg to adult, are totally planktonic. Recently, however, Dr Rees has studied the distribution of the larvae of the Decapoda—crab-like animals which live on the sea bottom. The larvae, quite different in appearance from the adult, are set free into the water and spend several weeks in a planktonic phase before they settle and change into the young bottom-living animal. The adult animals, benthic as they are called, are of great importance as food for bottom-feeding fish, and the planktonic larvae are often an important constituent of the food of pelagic fishes. Figure 16 gives the pooled numbers of total larvae on a number of recorder lines for several successive years. Although there is some variation in total production from year to year, it is clear that the pattern of distribution is strikingly constant; the larvae are, to a large extent, restricted to the southern part of the North Sea. In this region the bottom conditions are favourable to the adults and indeed it is known that there they are most common. Young larvae would be expected where the adults are most abundant, but the recorder survey also shows that they are retained in the same region throughout their planktonic life. Why are the larvae not more widely distributed? The restrictive factor to their northward spread is probably in part due to the barrier of the Dogger Bank and in part to a swirl of water in the Heligoland Bight, where the larvae circulate in a large eddy and are not carried away to the north. We may also note the scarcity of larvae in 1946 and 1947; on examining the temperature conditions recorded for this region it was found that the low bottom temperatures, which are always established during the winter in the North Sea, were maintained well into the summer of 1947. This is unusual and it was suggested that these low summer temperatures had a detrimental effect on larval production. Perhaps it is worth adding that the maintenance of these low temperatures was probably due to a relatively small influx of Atlantic water into the North Sea during that year.

Valuable results have been obtained by this technique and the work, originally confined to the North Sea, has now been extended to include parts of the Atlantic Ocean. However, this expansion creates new problems—the estimated number of recorder miles per year is 30,000, that is, 3,000 10-mile blocks— what a wealth of potential information! But how is a staff to keep pace with the analysis of all these records? Again it is a question of deciding on an objective and adjusting the working

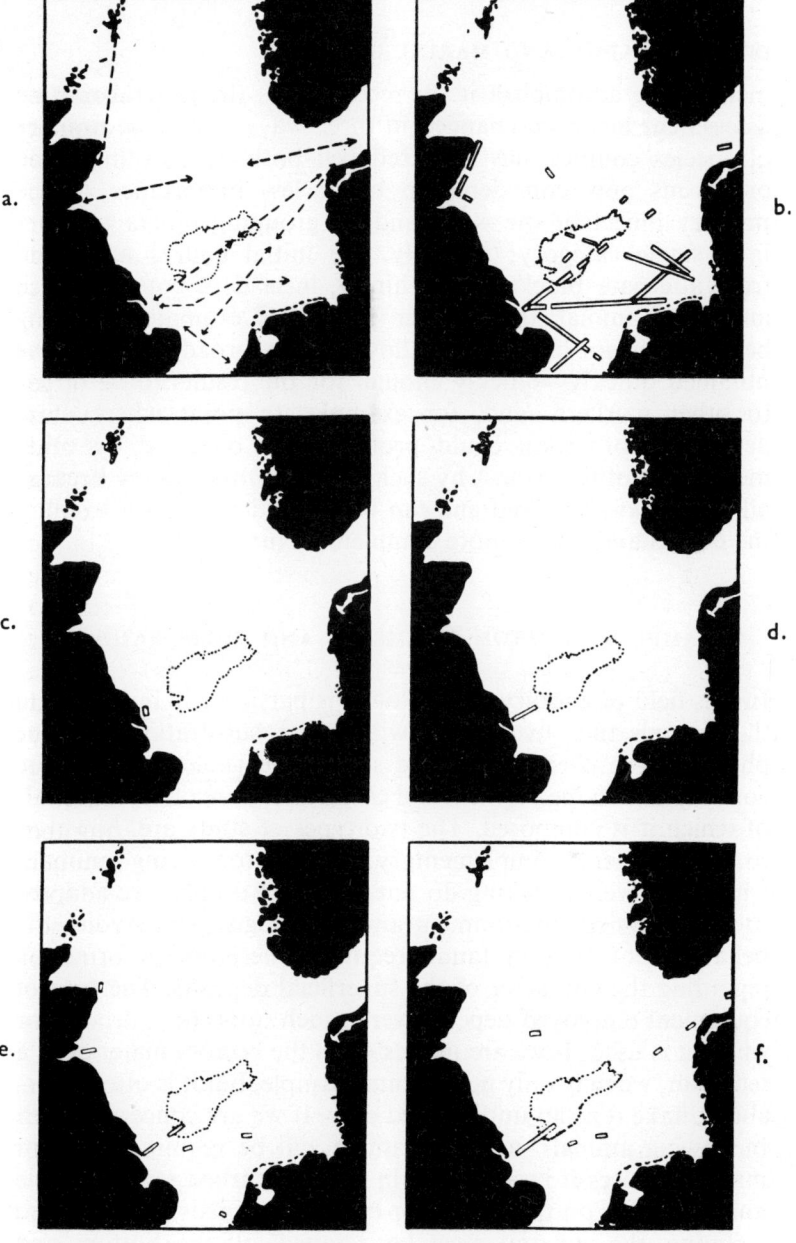

Figure 16
Occurrence of decapod larvae—Southern North Sea
a) The ship routes on which recorders were towed.   *b*) larvae 1938-9.
c) larvae 1946.   *d*) larvae 1947.   *e*) larvae 1948.   *f*) larvae 1949
The similarity in the distribution in these years and the lack of any wide dispersal is clearly indicated
(Dogger Bank dotted)
(*C. B. Rees*, '*Hull Bulletins of Marine Ecology*', 1952; *Modified from original*)

methods to accomplish it. In recent years the general routine assessment has been changed in three ways. First, the number of species counted has been reduced both by omitting some organisms now considered to be of less importance to the primary object of the work and by grouping others together in a single category; secondly, the initial counts are taken from alternate blocks only; thirdly, instead of counting each individual animal, the number present is estimated as lying between certain limits. This allows a very broad picture to be obtained quickly—quickly enough for the results to be of use to other workers—and, for example, to point where more detailed information could profitably be obtained by other methods. Nothing is lost by such a rapid initial survey because all the blocks are stored and can be worked over again to fill in the details and give a more complete picture.

SAMPLING BOTTOM SEDIMENTS AND THEIR FAUNA

In this field of enquiry the biologist is particularly interested in the animals that live on or within the substratum, but the physicist, the chemist and the submarine geologist are more concerned with the physical and chemical nature of the materials of which it is composed. The two types of study are, however, to some extent complementary; the bottom-living animals, which relatively speaking do not penetrate deeply, are adapted to their physical environment and a biologist who studies the behaviour of bottom fauna requires, therefore, information regarding the character of the superficial deposits. The type of equipment employed depends very much upon the purpose for which it is used. If we are interested in the bottom material as a sediment, we may only need a small sample, but it is often desirable to take it as an undisturbed core. If we are concerned with burrowing animals, a larger sample will be required and for many purposes it need not be in an undisturbed state. For the animals living on the surface or on other animals, the so-called epifauna, the material must be 'scraped' off the bottom and when the animals are relatively sparsely distributed a consider-

SAMPLING THE LIVING ORGANISMS

able amount of ground may have to be covered. All these objectives require different sampling devices each adapted for its own specific purpose; none is adequate for all.

Trawls and dredges

A commercial trawl is used for sampling bottom-living fish; the foot-rope, often rigged with special devices such as tickler chains, disturbs the fish, which pass backwards into the cod-end as the trawl is dragged through the water.

For many purposes the commercial otter trawl, which cannot be easily worked from a small boat, is too large and the meshes too wide for sampling organisms other than fish. A beam trawl (Figure 17), now little used commercially except in some inshore prawn fisheries, may, however, be employed. The rigid front of such a trawl consists of a stout wooden beam fastened between two metal shoes, the soles of which slide over the ground. The upper edge of the net is fastened along the beam and the sides to the shoes. The lower, leading-edge of the net is attached to a stout foot-rope and is shaped so that in towing it lies in a curve behind the front of the net. Commonly, the fore part of the net is made with a coarser mesh than the after-collecting portion. The trawl is towed on a single warp fastened to a pair of bridles, the latter being shackled to eyes on the metal shoes. When towed, the length of warp out must be adjusted both to the speed of the ship and to the depth of water so that, on the one hand, the trawl does not dig into soft ground and fill with mud, while on the other it does not travel too lightly over the bottom and fail to take off the animals.

The Agassiz trawl is used for the same purpose and is similar in construction but the shoes, connected between their centres by a stout iron bar, are symmetrical and there is no beam. It has the advantage that since it is completely symmetrical with regard to its upper and lower parts it will tow and catch material whichever way it falls on the ground.

On fairly soft and even ground—mud, sand or gravel—the beam or Agassiz trawl is a suitable device for taking the

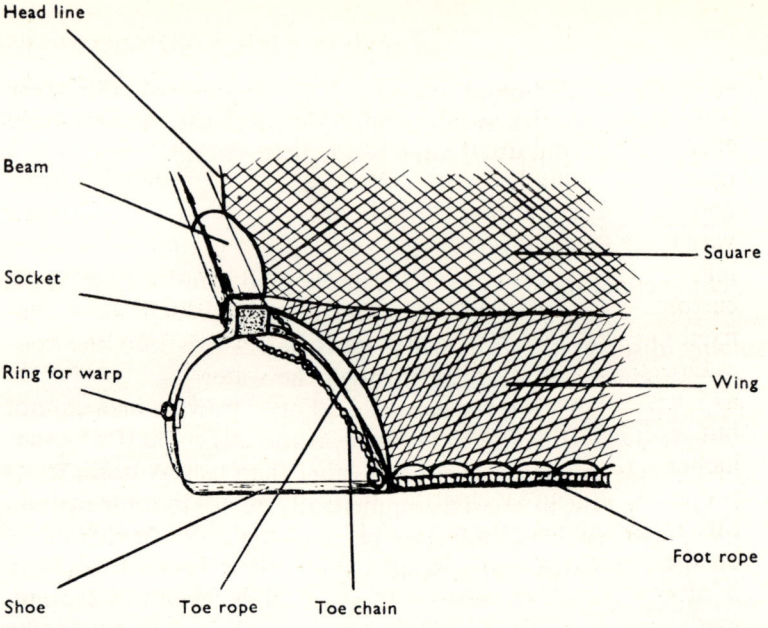

Figure 17
a) Beam trawl; side view showing metal shoes, beam and net

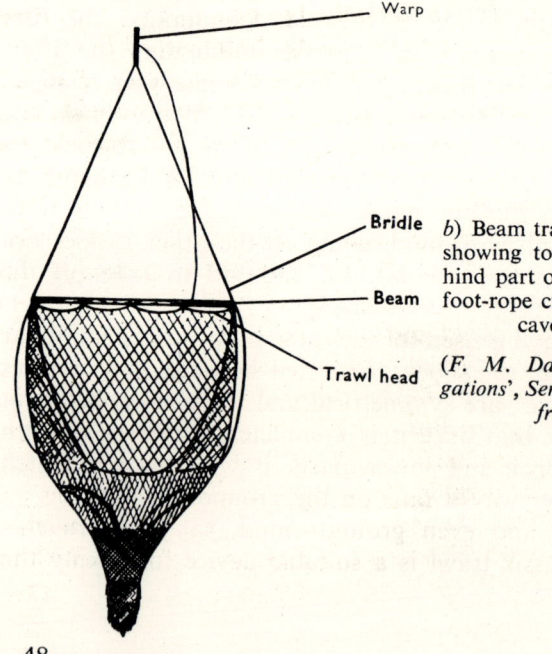

b) Beam trawl; view from above showing towing warp, fore and hind part of net. Note how the foot-rope curves to form a concave leading edge

(F. M. Davis, 'Fishery Investigations', Series II, 1927; Redrawn from original)

SAMPLING THE LIVING ORGANISMS

epifauna. On rough, stony or rocky ground, the net is often torn either by the substratum itself or by large stones that enter the net and are dragged over the bottom. Sometimes the trawl may 'come fast' and be lost. On such grounds a stronger and smaller apparatus, the dredge, is used. This may take a variety of forms often adapted to local conditions and usually influenced by personal preference or local prejudice and custom. It is much smaller than a trawl and usually has a fixed and strong metal opening to which the net is attached.

Dredges are used commercially for harvesting shell fish such as cockles. The lower edge of the frame may be a simple bar, but more often it has a blade-like form with the leading-edge inclined downwards and forwards; it may have downward-projecting spikes fastened to it. In either case the function of this modified leading edge is to scoop up animals lying free on the bottom so that they pass backwards into the bag which is attached to the front frame. The floor of the bag is often made from a series of large iron rings—rather like a piece of chain mail—whilst the sides and roof are of net. Care must again be taken to adjust the length of warp and the speed of tow to the nature of the ground so that the dredge neither digs too deeply into soft ground nor passes too lightly over hard ground.

For sampling burrowing animals some form of dredge is used; they are often made with a very heavy rigid metal frame, oval in shape, and to this the net is attached. In use, they are allowed to take a sharp bite into soft ground and bring up a 'bag' of mud.

On sand, where there is considerable resistance to the leading-edge of either the mud bucket or naturalist's oval dredge, only a small sample is usually brought up and deeper burrowing animals frequently avoid capture. To get over this difficulty Mr Forster of the Plymouth Laboratory has devised what he calls an anchor dredge, which is shot and hauled in a similar manner to a ship's anchor. The net is attached to a strong rectangular frame on the lower bar of which fits a stout iron base-plate with a digging leading-edge. A strong iron bar projects forward from the centre of the upper edge of the frame and to this the warp is shackled. The dredge is put out with the ship going astern. A long length of warp is then paid off; on hauling, the broad inclined plate digs into the substratum and a large and deep bite is taken. The long forward-projecting arm serves to prevent fast-moving burrowers from being disturbed by the warp and making their escape as the sample is taken; with this instrument many deep-burrowing animals are taken that usually avoid capture by more conventional dredges.

Grabs

If we wish to investigate the ground quantitatively, then an attempt must always be made to take a sample of equal area and volume. Some form of grab is used for this purpose, but it is extremely difficult to devise an instrument which will work adequately under all circumstances.

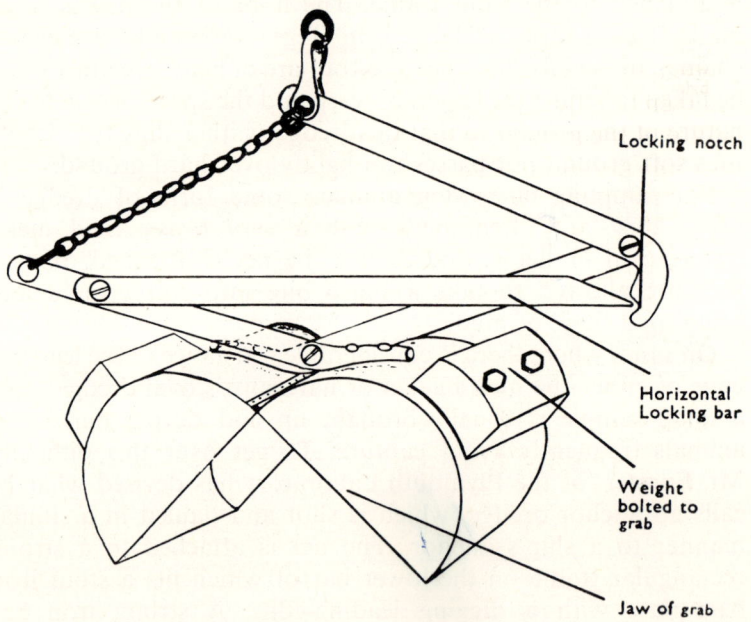

Figure 18

The Petersen-type grab

SAMPLING THE LIVING ORGANISMS

In the Petersen (Figure 18) or van Veen type of grab two semicircular buckets are hinged along a central axis. The buckets are held apart by some form of catch. On striking bottom this is released, so that on hauling the buckets move round on their axis, take a bite out of the ground, and eventually come together to form a closed container.

Neither of these grabs is invariably successful on all types of ground and under all working conditions. The rate at which the grab hits the bottom affects the bite and when the ship is drifting, a poor sample may be obtained if the grab does not hit the ground vertically. Further, the depth of penetration varies with the type of ground; material may be lost if shells or stones lodge in the jaws and prevent complete closure.

A number of attempts have been made to devise an instrument in which these faults are absent. In that designed

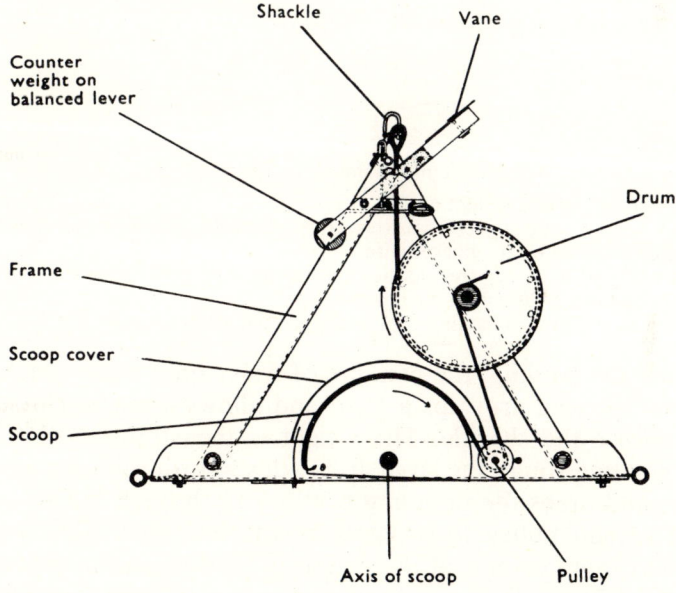

Figure 19
The Holme mud-sampler with semi-circular scoop
(*N. A. Holme, 'Journal of the Marine Biological Association, U.K.'*, 1949; Redrawn from original)

51

by Mr Holme of the Plymouth Laboratory the sample is taken by a scoop rotating on an axle mounted on a heavy frame that rests firmly on the ground (Figure 19). (Single and double scoop models have been designed.) The apparatus is lowered open (Figure 20), the whole weight being taken by a shackle on a balanced arm; closure is not effected on touching

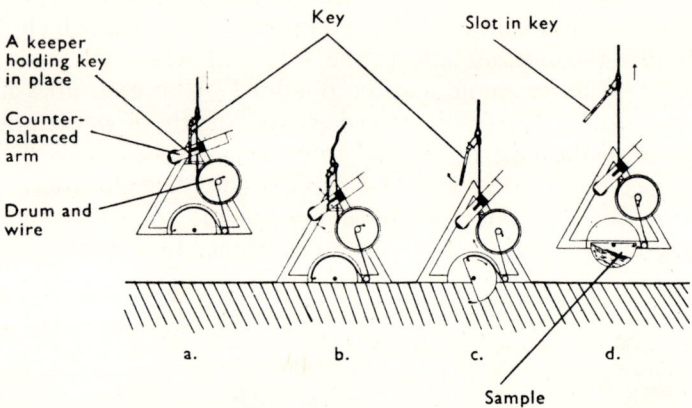

Figure 20

The Holme mud-sampler (diagrammatic)

a) Set with pin in key, weight on key
b) Arm tips on touching bottom when warp slackens, pin slips out allowing tension to come on the bucket wire
c) Beginning to haul, key free, tension on hauling wire now comes on wire on drum and this pulls round the scoop
d) Scoop now round as hauling continues and completely closed

bottom. On hauling, the pressure of water on a vane attached to this balanced arm tips a lever and allows a pin to slip out and release the shackle. The weight is then transferred to a rope rotating round the large drum; this in turn rotates a small pulley and drags the sampling hemisphere through the bottom via a second pulley to which it is attached by a light wire. The maximum volume of the sample is $5\frac{1}{2}$ litres and in practice about 3-4 litres of soil are brought up; this is some four times that usually obtained by a Petersen grab ostensibly sampling the same surface area of $\frac{1}{10}$ square metre; again, it digs deeper and brings up species that are not normally taken by the older grabs.

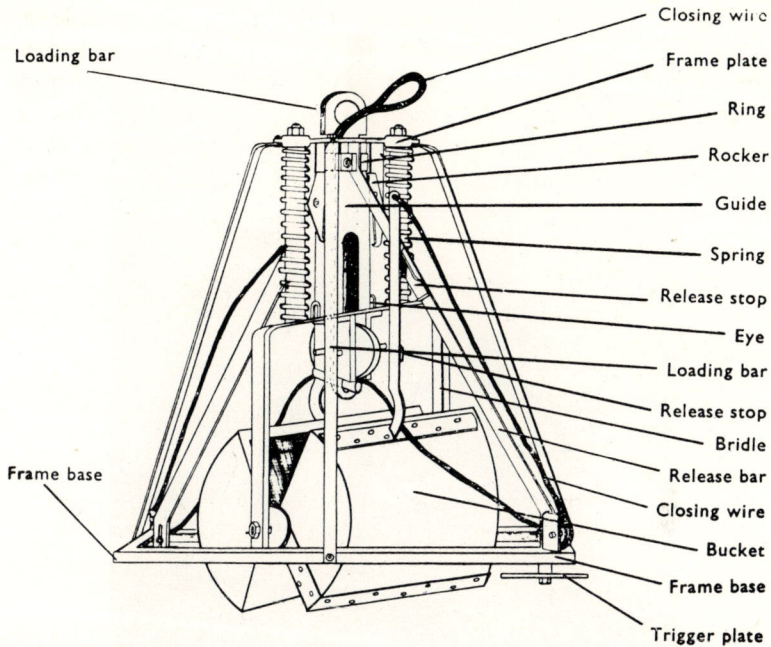

Figure 21
The Smith-McIntyre spring-loaded mud-sampler. In this instrument heavy springs force the grab into the soil before the jaws are closed to collect the sample
(*W. Smith and A. D. McIntyre*, '*Journal of the Marine Biological Association, U.K.*', 1954; Redrawn from original)

The van Veen, Petersen and Holme samplers do not work very satisfactorily in poor weather. Messrs Smith and McIntyre have described a new sampler (Figures 21-23) in which the digging and hauling mechanisms are separated and whose performance is good even under bad working conditions. It consists of a spring-loaded bucket actuated only after the apparatus rests squarely on the bottom (Figures 21 and 23). The bucket is carried on a stout metal bridle which slides on a guide tube, and two heavy springs on the frame bear on the movable bridle. In the 'set' position the springs are held by two catches which fit into a pair of holders and these are triggered when the apparatus strikes bottom. On release, the springs bear on the bridle and push the bucket down into the substratum. At the

Figure 22
Smith-McIntyre grab
Apparatus set and going over
(*From W. Smith and A. D. McIntyre. 'Journal of the Marine Biological Association, U.K.', 1954*)

same time as these are triggered, a release bar moves so that on hauling the weight comes on the wire and the two halves of the bucket are drawn together taking a sample of soil.

All the grabs so far described are inefficient on very sandy bottoms; this is in part the result of the great resistance of sand to their digging action, and in part to the difficulty of keeping coarser samples intact during hauling. A vacuum grab such as described by Holme may then be used (Figure 24).

In this instrument the force required to suck in a sample of the bottom is provided by a 'vacuum' chamber which contains

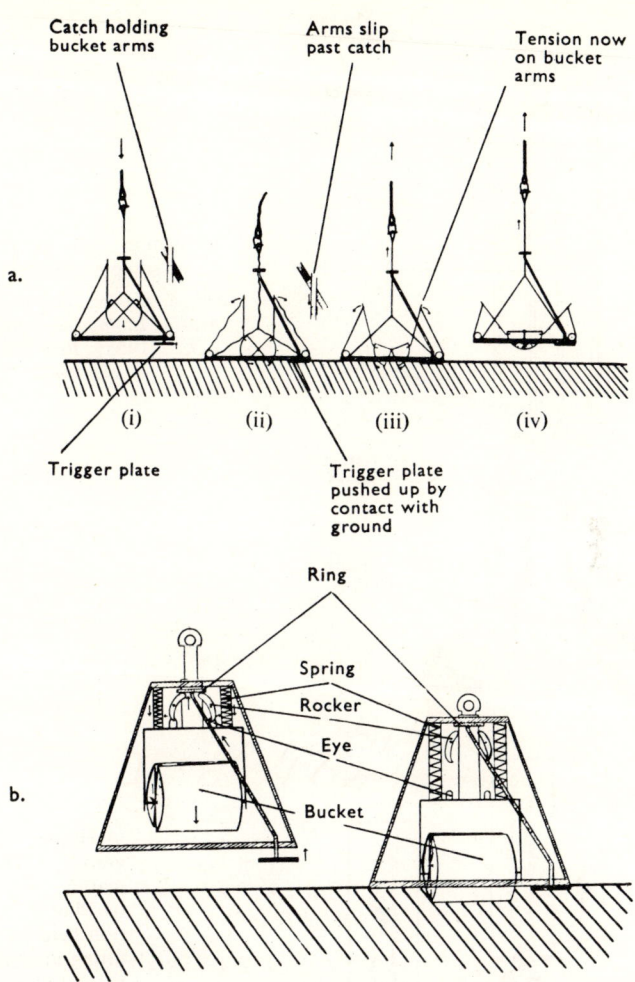

Figure 23

The Smith-McIntyre mud-sampler (diagrammatic)

a) The bucket mechanism
  (i) Descending
  (ii) Hitting the ground—cable slackens, trigger plate allows arm to slide past stop
  (iii) Hauling begins, weight on arms close the bucket
  (iv) Further hauling, bucket completely closed

b) The spring-loading device
  (i) Springs set and bucket held by rockers in eyes and by ring
  (ii) Trigger plate thrust up, ring moves up, allowing rockers to slip out of eyes and springs, then thrust bucket into the mud

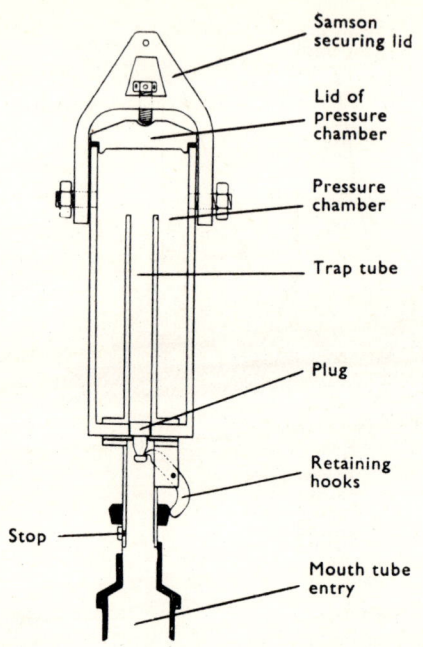

a) Suction type of bottom sampler (Holme)
(*N. A. Holme*, '*Journal of the Marine Biological Association, U.K.*', 1955; Redrawn from original)

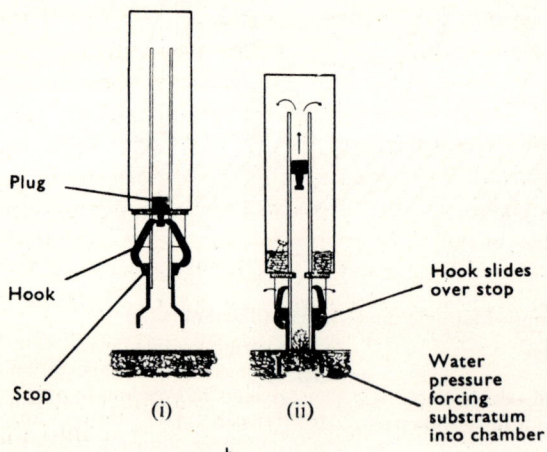

b) Holme suction grab (diagrammatic)
(i) Plug in set position held by hooks fitting into slots
(ii) Released by bottom of hooks sliding over stops on contact with the bottom, plug flies up and mud is sucked into pressure chamber

Figure 24

SAMPLING THE LIVING ORGANISMS

air at atmospheric pressure and is mounted above the collecting tube. On striking the bottom the chamber is put into communication with the outside and the water pressure forces the sample into the collecting tube.

The pressure chamber itself is a strong brass tube closed at the upper end by a lid and held firmly in position by a clamp. A sampling tube is fixed below the chamber and extends upwards into it. Between the upper and lower parts of the central tube is a plug held in position by retaining hooks through slots in the wall of the lower tube. When the apparatus strikes the bottom a mouth tube rises, disengaging the plug, which flies up the tube; the water pressure then forces material up into the collecting tube (Figure 24). The lid is removed and the sample emptied out.

Coring devices for undisturbed samples

For many purposes it is sufficient to be able to take a sample with one of these instruments and examine the animals contained in it. For some types of work a sample of bottom material remaining as it is *in situ* is needed; this is often the case with mud samples and for this purpose a coring device is used. This takes many forms; in all cases the instrument must have some simple type of valve that lets water pass through the apparatus during descent but which closes when the corer is raised to prevent loss of the sample during hauling. The complexity of this type of device largely depends upon the length of core required. In studies of the relation of burrowing animals to the material in which they live, or for investigations of recent sediments, a core of superficial material a few inches to a foot is adequate and the apparatus is relatively simple. For geological purposes, the longer the core the greater is the yield of information and a more complex and heavy instrument is essential.

For short cores, the apparatus is variously modified; it may be a long tube with attached weights which drive it into the bottom and which are then released, or the weight of the sampler itself may be sufficient for this purpose. Moore and Neill's sampler is very effective and will be described (Figure 25). It is essentially a protected glass tube through which water flows freely during descent and which is forced by impact into the mud; on hauling, a simple valve mechanism closes the top of

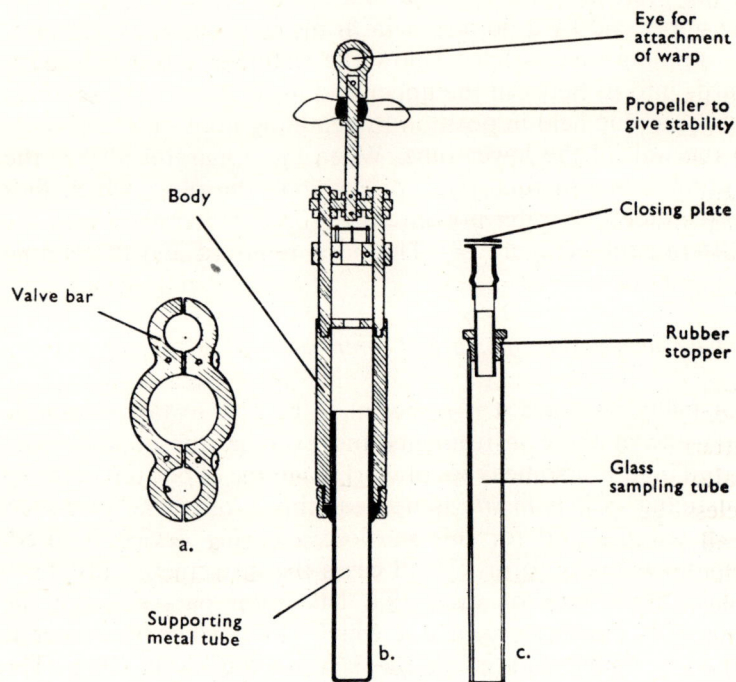

Figure 25
Coring device for short cores
(*H. B. Moore and R. G. Neill*, '*Journal of the Marine Biological Association*', U.K., 1929-30; *Redrawn from original*)

the tube and the sample may be brought to the surface. The body of the sampler is a stout brass cylinder (Figure 25b) and into this fits a thinner metal tube holding the glass sampling tube (Figure 25c). When the glass tube is in position it is closed by a rubber bung which comes hard against the upper surface of the main body of the sampler in the centre of which is a hole. Through this passes a short piece of small-bore glass tube

SAMPLING THE LIVING ORGANISMS

which in turn passes through the rubber bung closing the glass sampling tube. The upper end of this short glass tube is joined by rubber to a second similar tube which fits against a simple valve. The latter consists of a flat ground rim on which lies a ground glass plate confined in a cage. As the sampler descends, water streams through the whole apparatus and escapes through the valve, since the glass closing-disc is lifted up in its cage by the upward pressure of the water. The weight of the sampler drives the tube into the bottom and a core of material is forced into the glass sampling tube; the glass plate then drops on to the ground surface (being no longer forced up by the water) and when the instrument is raised gives a watertight joint. At the upper part of the apparatus is a stout pillar on which is mounted a propeller revolving freely and thereby helping to maintain a vertical descent.

We have seen how, in several branches of oceanography, rapid methods of survey are of great importance, particularly when we want a synoptic picture of a continually changing pattern. In the case of surface sediments, except in the most shallow waters, the pattern is relatively stable; it may, nevertheless, be of a complex character with sharp boundaries over small distances. If the time taken to get a sample can be reduced, then it becomes economically and practically possible to carry out more detailed surveys of superficial bottom deposits.

To this end Emery and Champion have designed and used an under-way sampler (Figures 26 and 27); with it they have taken many thousands of samples—as many as 200 in a working day. The instrument consists of a sampling cup which is mounted in a hollow barrel attached to a tail-piece. When cocked ready for use, the cup protrudes about half an inch from the barrel and a hinged cover is held back by a spring. Attachment to the light towing cable is by means of a hinged arm, that is folded down during descent (Figure 26b). When ready, the sampler is gently lowered over the ship's side, but on striking the water the brake is taken off the winch drum and the cable rapidly run out. On striking bottom, a small sample is taken into the cup, which at the same time is forced back

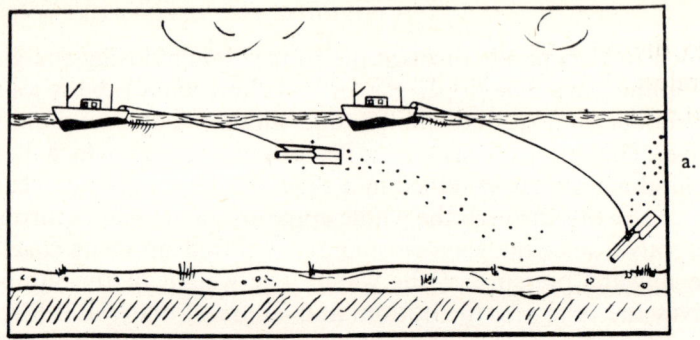

Under-way bottom sampler

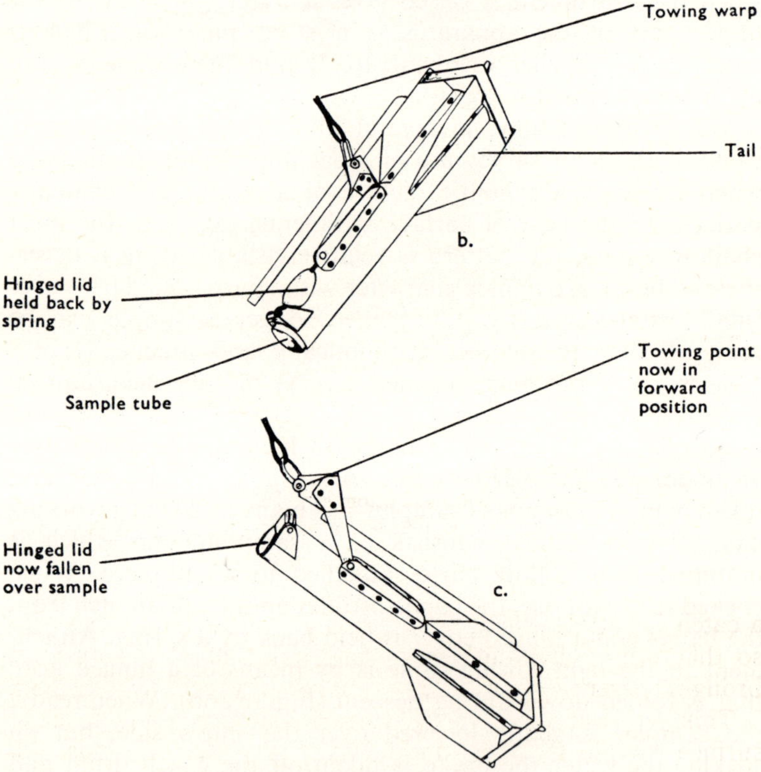

Top, going down; *bottom*, coming up

Figure 26

SAMPLING THE LIVING ORGANISMS

into the tube; as this happens, the spring holding back the lid is released and the latter now comes over the end of the tube and prevents any loss of sample during hauling. Simultaneously

Figure 27
Under-way bottom sampler;
ready for shooting
(*Courtesy of Professor K. O. Emery*)

a catch is sprung which frees the after end of the towing arm so that during hauling the towing point of the sampler is brought nearer the forward end, thus reducing drag (Figure 26c).

This technique has some severe limitations. Only a small surface sample is taken. In the case of muds or sands, the amount is usually sufficient for standard chemical and physical analyses, but with gravels and coarser materials only a few fragments are often brought up; even so, these may be sufficient to characterize the bottom. Really coarse material, such as moderate-sized stones, is not taken.

The apparatus has been very successfully used to make rapid surveys in water of depths up to 100 fathoms—changes of surface sediments being correlated with changes in topography as revealed by echo sounding. Clearly the method may be extended and developed to take larger samples and larger-sized material from the bottom; it has not been used for biological work.

We may now consider the specialized corers used for geological work.

The older coring devices were modifications of the weighted tube; success depended upon their weight and the speed at which they could be dropped. Much valuable information from cores 3-4 ft. long was obtained by the German *Meteor* Expedition, using such instruments developed by F. L. Ekman and improved by V. W. Ekman. A corer developed by C. S. Piggot gave cores up to 10 ft. long; this instrument did not rely upon weight or speed of fall, but was driven into the substratum by an explosive charge. Although the long cores taken by this instrument gave a great stimulus to coring studies it is not much used at present, since cores of similar length can be obtained by newer corers that are much cheaper to make and safer to use. The Kullenberg corer has revolutionized this field; it will take cores up to 70 ft. in length. This and the Emery-Dietz gravity corer will be described.

In the latter corer an attempt has been made to produce a cheap, robust and reliable instrument capable of working even under moderately adverse sea conditions (Figures 28, 29, 30). It consists essentially of a shaft, weights, and coring tube. The shaft is a standard 8-ft. pipe $2\frac{1}{2}$ in. in diameter with several small holes drilled in it to allow water to run off as it is lifted on board. On this shaft is mounted a set of removable lead weights, which are roughly streamlined; the number of weights is adjusted to give the required depth of penetration. The complete corer weighs some 650 lb. in air. The weights rest on a shoulder, which is part of the valve housing. The valve is similar in principle to

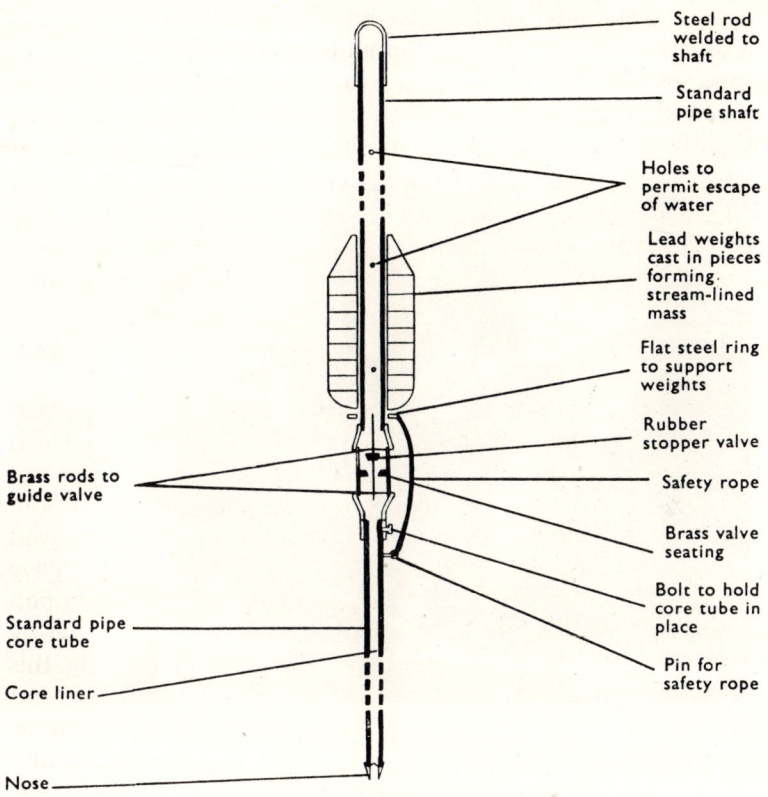

Figure 28
Emery-Dietz gravity corer
(K. O. Emery and R. S. Dietz, 'Bulletin of the Geological Society of America', 1941; Redrawn from original)

the ground-glass plate of the Moore and Neill sampler and is necessary to prevent loss of the core by suction during withdrawal from the mud, to obviate any washing out of the relatively fluid upper part of the core during hauling, and to stop the weight of water in the shaft forcing out a short core as the instrument is lifted out of the sea. It consists of a rubber bung resting in guides, and fitting loosely in a brass seating. The coring tube itself fits below this valve and consists of a standard pipe, 2 or $2\frac{1}{2}$ in. in diameter, thrust into a sleeve against a shoulder of the valve housing and clamped there with a set-screw. A removable celluloid liner is fitted inside the coring tube so that the core may be readily removed and inspected. Screwed to the bottom end of the tube is a nose-piece with a slightly smaller diameter than the coring tube itself. A core

Figure 29
    Emery-Dietz gravity corer being brought on board R.V. *Velero IV*
    (*Courtesy of Professor K. O. Emery*)

retainer is fitted in the nose; this has strips of celluloid thrust into a thin rubber sleeve and bent over towards the centre. When the sample is passing up the coring tube they are pushed back against its wall, but if the core starts to slip, they close the opening. In use, the instrument is lowered on a $\frac{3}{8}$-in. wire until about 300 ft. above the bottom when it is allowed to run almost free with only sufficient braking to ensure a vertical descent. Just before striking the bottom the velocity is between 12 and 21 ft. per second. After hauling and hoisting inboard (Figure 29), the undisturbed core is extracted from the tube by withdrawing the celluloid liner; the core is then cut into convenient lengths which are labelled and stored in sealed containers.

Figure 30

Washing a mud sample on board R.V. *Velero IV*

(*Courtesy of Professor K. O. Emery*)

If it is not possible to have a free-running winch, the method suggested by Hvorslev and Stetson may be used (Figure 31). In this the corer is attached by a coil of wire to a release mechanism from one arm of which hangs a counter-weight. This weight, hanging the required distance below the corer,

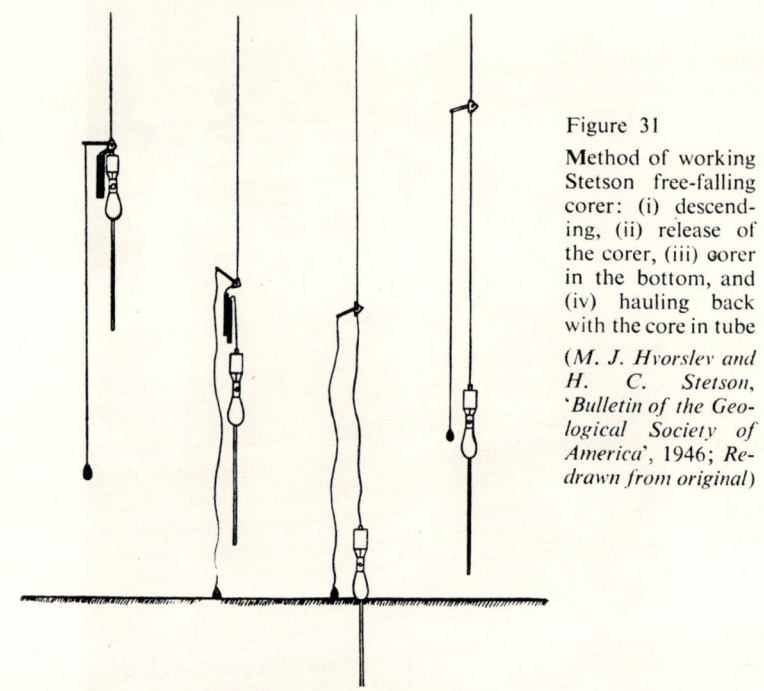

Figure 31

Method of working Stetson free-falling corer: (i) descending, (ii) release of the corer, (iii) corer in the bottom, and (iv) hauling back with the core in tube

(*M. J. Hvorslev and H. C. Stetson*, '*Bulletin of the Geological Society of America*', 1946; Redrawn from original)

touches bottom first and allows the corer to slip off the release and then to fall under its own weight.

The length of a core taken with an instrument driven into a sediment, either by its own weight and velocity or by an explosive charge, is always less than the depth of penetration of the coring tube. Depending upon the instrument and the type of sediment, the core itself may vary from 40 to 70 per cent. of the depth of penetration. To what extent then does a point on the core correspond to the original sediment? Does

SAMPLING THE LIVING ORGANISMS

material escape from the tube mouth, or are the layers simply compressed into a shorter length? The answer to these questions is of fundamental importance in any interpretation of cores, particularly when attempts are being made to estimate rates of sedimentation of the layers. A good deal of attention has been paid to this problem. The main factor responsible for shortening is friction between the core, as it passes up the tube, and the walls of the latter. As more and more sediment passes up the tube, frictional resistance increases and it becomes more and more difficult for material to enter; only part of the sediment does so, the rest slipping away so that the layers in the tube become increasingly smaller than those in the undisturbed sediment. Eventually frictional forces become so great that they prevent any more sediment entering the tube even though the latter continues to penetrate the soil. (Pratje reduced this friction by fitting a nose-piece with a slightly smaller diameter than the tube and so greatly increased the length of the cores taken by Ekman-type instruments.)

In Kullenberg's piston corer the hydrostatic pressure at great depths is utilized to overcome this friction and to allow very long undisturbed cores to be taken (Figures 32, 33 and 34).

Consider a tube (Figure 32a) being forced into a sediment, and suppose a closely fitting piston is mounted in the tube just above the surface of the sediment. As the tube goes down, wall friction exerts a downward pull on the core; if this were to let a space develop between the piston and the surface, a partial vacuum would be set up. Hydrostatic pressure, when it exceeds frictional forces (which will always be the case in deep water), will not allow this, and an undisturbed core is drawn into the tube. The key to the problem is, therefore, to immobilize a piston inside the tube in the vicinity of the bottom. This has been done ingeniously in Kullenberg's instrument—which has revolutionized coring techniques and sediment studies.

The sampler consists of a hard steel tube which may be up to 70 ft. long loaded with weights according to the length of core required and the consistency of the sediments. To avoid damaging the whole tube should it strike rock, it is made in 15-ft. sections coupled together. At its lower end is a nose;

this has a slightly larger external and slightly smaller internal diameter than the tube itself. The larger external diameter reduces friction on the outside of the tube by the water descending in its wake. The smaller internal diameter is not required in the piston sampler to obtain longer cores by reducing

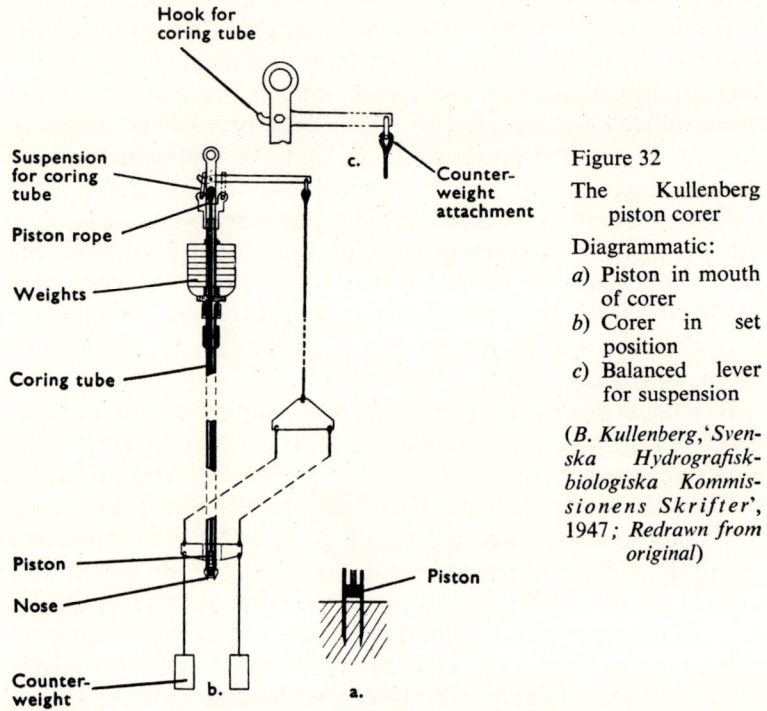

Figure 32
The Kullenberg piston corer
Diagrammatic:
a) Piston in mouth of corer
b) Corer in set position
c) Balanced lever for suspension

(B. Kullenberg,'Svenska Hydrografiskbiologiska Kommissionens Skrifter', 1947; Redrawn from original)

the internal wall friction, but a reduction in this is still desirable to prevent distorting the peripheral parts of the sample. The steel tube is lined with short lengths of brass tubing. The piston fits the brass liners and is connected by a steel rope to the suspension. The corer is suspended from the shorter arm of the release mechanism, the longer arm of which carries paired heavy counter-weights. These weights are suspended through eyes in a metal bar carried by the main tube. They hang down symmetrically below the mouth of the coring tube so that the

SAMPLING THE LIVING ORGANISMS

Figure 33
The Kullenberg piston corer
The corer is ready for lowering; the safety catch has been removed and the counter-weights hold the corer in set position
(*Courtesy of Professor B. Kullenberg*)

centre of gravity is located on the axis of the tube, which is therefore kept in a vertical position. The counter-weight rope is used to bring the apparatus inboard (Figure 32).

The counter-weights strike the bottom first and unload the release mechanism, allowing the lever to tip and the corer to slip off the shorter arm and fall into the sediment with a considerable velocity. As the tube slips off the tension in the cable is reduced and this is indicated by a ship-board dynamo-

Figure 34

The Kullenberg piston corer
(*Courtesy of Professor B. Kullenberg*)

SAMPLING THE LIVING ORGANISMS

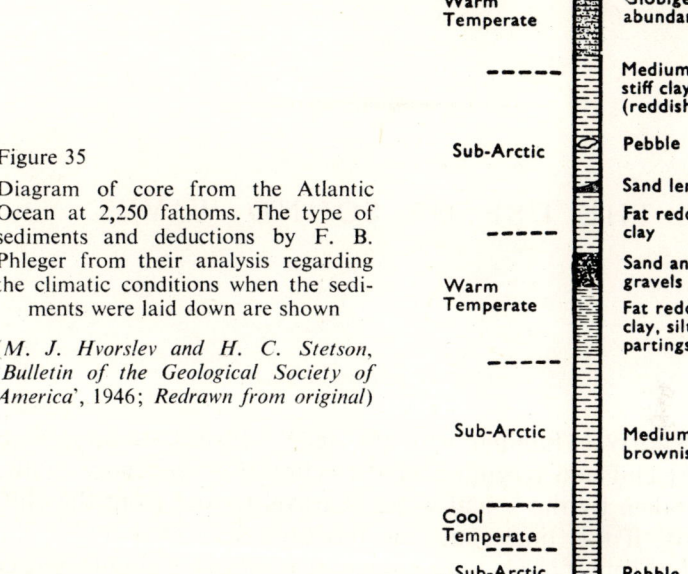

Figure 35

Diagram of core from the Atlantic Ocean at 2,250 fathoms. The type of sediments and deductions by F. B. Phleger from their analysis regarding the climatic conditions when the sediments were laid down are shown

(*M. J. Hvorslev and H. C. Stetson*, 'Bulletin of the Geological Society of America', 1946; Redrawn from original)

meter; the winch is stopped and the piston arrested above the bottom, in which position it stays as the core enters the tube.

In Figure 35 is a diagram of a long core, taken by Hvorslev and Stetson, in which the types of sediment are indicated together with the associated climatic conditions prevailing when these components of the core were laid down. The climate is inferred from the present known distribution of the organisms whose remains are found in the core when it is submitted to detailed analyses.

71

2

THE USE OF SOUND WAVES

THE ECHO SOUNDER

EVERYBODY is familiar with the delay between shouting at a distant cliff and reception of the echo—a consequence of the time taken by the sound wave to travel to and from the cliff. Clearly, if this time could be measured accurately and if the speed of sound in air were known, it would be a simple matter to calculate the distance away of the cliff.

The speed of sound in water is known (about 4,900 ft. per second) and it is this simple principle of measuring the time taken for an echo to return from the sea bed that is used in all echo sounding (Figure 36)—a method of depth recording which has almost completely replaced the older and more laborious lead line. The suggestion that sound waves should be used in this way is usually attributed to Jean F. Arrago about the year 1807. At first submarine bells, explosions, and hammers, were used to generate audible sound waves (relatively low frequency), and the time taken by these to return was measured as accurately as possible by a chronometer. The microphone was introduced about 1900 as a sensitive means of detecting the return echo. Echo sounding was not, however, considered sufficiently reliable for surveying purposes until 1924, in which year it made its first appearance in the Hydrographic Service of the British Navy. Rapid development followed and the

THE USE OF SOUND WAVES

reliance placed on the new method may be judged by the fact that in 1929 it was used for some 60 per cent. of all the soundings made by the Navy.

In early machines sound waves, at audio-frequency, were produced by an electromagnetic hammer and the echo picked up by hydrophones. The transmission was timed by a bridge control unit which also passed the echo, via a transformer, to telephones. A telephone short-circuiting device coupled to a movable depth scale allowed the echo, detected by a 'click' in the telephones, to be located in relation to a scale reading. Although now quite out of date, such sets were used for charting depths up to 4,500 fathoms.

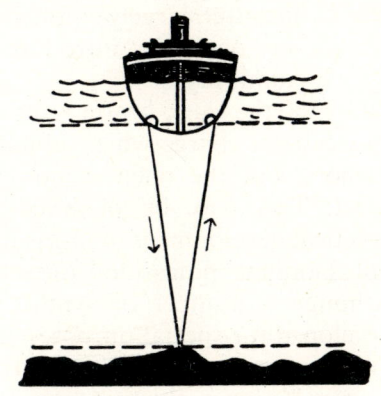

Figure 36

The principle of echo sounding. The time taken for an echo to return from the sea bed to the ship is timed, and the distance 'calculated' from the known speed of sound in water

The production of ultrasonic sound waves

The use of low-frequency audible sound has two inherent disadvantages: first, it is difficult to separate the signal from any nearby noise such as that coming from the ship's engines and propellers, and secondly, it is difficult to screen the receiver from the transmitter. An important advance was made when audio-frequency transmitters were replaced by those giving an ultrasonic beam, that is, working with sound of a frequency just at or higher than can be detected by the human ear. The sound frequency used in modern machines varies from 15 to about 50 kilocycles per second.

These ultrasonic high-frequency waves are produced in two ways, namely, by utilizing either the piezo-electric or the magneto-striction effect. The former was used by Professor Langevin, working for the French Government on the development of methods for detection of underwater craft during World War I; the latter, largely employed in British commercial instruments, was originally introduced by Dr A. B. Wood of the British Admiralty.

In 1880 the Curie brothers found that certain crystals develop an electrical charge when submitted to mechanical pressure or tension, and the phenomenon was termed the piezo-electric effect. This may be observed with many crystals; for the practical development of ultrasonic beams, quartz and Rochelle salt (sodium potassium tartrate) are most generally used, although a number of synthetic crystals have recently been developed for special purposes. The electrical charge developed on the crystal is proportional to the pressure applied and when a compression is changed to an equal tension, the sign, but not the intensity, of the charge changes. Conversely, when a charge is applied to one of these crystals, its faces move with respect to each other. If, therefore, such a crystal (and particularly when it is cut to oscillate at a given frequency) is submitted to an alternating high-frequency electrical charge and one surface of the crystal is then applied to that of a transmitting medium, the vibrations of the crystal will produce ultrasonic waves in that medium. It is in this way that the piezo-electric effect is utilized to generate ultrasonic high-frequency sound waves. Further, by suitable mounting of the crystals the mechanical changes that take place when sound waves fall on them and which in turn give rise to electrical effects can be employed to detect the return echo. In practice, the crystals are mounted in mosaics or stacked to give a greater output of sound energy.

The magneto-striction effect depends upon the fact that when certain metals, notably nickel, are placed in a varying magnetic field they undergo mechanical changes; nickel itself contracts in an increasing and expands in a decreasing magnetic field. As with the mechanical changes brought about by the piezo-electric effect, these also may be transmitted to a medium and, if they are sufficiently rapid, ultrasonic sound waves are produced. By altering the rate at which the magnetic field

THE USE OF SOUND WAVES

changes, the frequency of the sound pulses can be varied. In practice, the magneto-striction oscillator is made up of a number of nickel laminations wound with a few turns of thick low resistance wire. This is connected to a condenser placed in circuit with a high-voltage generating unit; when this condenser discharges, a momentary large surge of current flows through the oscillator winding resulting in the rapid formation and dying away of an intense magnetic field; the inductance of the oscillator winding coupled with the capacity of the transmitting condenser give an oscillatory circuit. This oscillatory magnetic field causes the nickel plates alternately to increase and decrease their diameter, that is, to vibrate radially, and when applied to a medium, ultrasonic waves are produced in it.

The modern echo sounder

We are now in a position to understand the working of a modern echo sounder; short pulses of high-frequency sound waves are sent out from the ship and the time required for the resulting echoes to return from the sea bed is measured; the records are usually presented in the form of a continuous graphical trace of the depth of water beneath the vessel. To use the principles already outlined in a convenient way, we require:

(*a*) an ultrasonic sound transmitter and receiver,
(*b*) a means of amplifying the return signal,
(*c*) some form of scale, calibrated directly in distance,
(*d*) a method of recording the distance indicated by this scale.

There are many types of instruments but all use the same principles. The general layout is shown in Figure 37, and a typical instrument, the Kelvin and Hughes M.S. 26 Survey Echo Sounder, in Figure 38. Most British instruments utilize magneto-striction transmitters and receivers. The transmitting unit or oscillator in wooden ships is of the pierced-hull type, a hole being made in the ship's hull and a plate inserted astride the oscillator; alternately it may be mounted in a separate external dome. When the plates of an iron ship are not too thick, the oscillator is fixed directly on the hull (Figure 39). The oscillator itself is contained in a tank filled with water and the radial vibrations of the nickel stampings, transmitted hori-

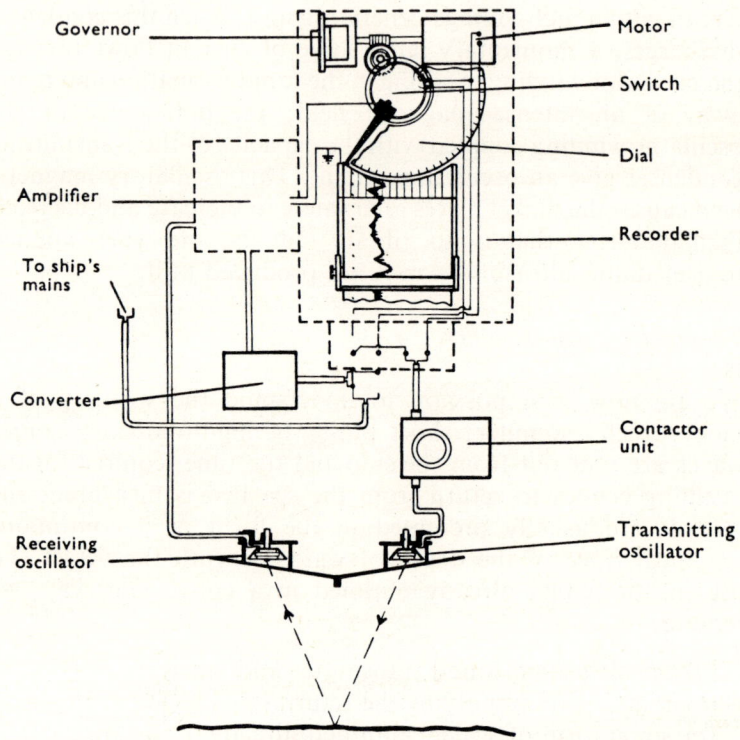

Figure 37
Diagram of Kelvin and Hughes-type echo sounder. The contactor box which generates the signals is connected to the oscillator. The return sound is received, passed to the amplifier and then to the stylus of the rotating arm which marks the paper
(*Courtesy of Messrs Kelvin & Hughes; Redrawn from original*)

zontally through the surrounding water, are condensed into a beam and projected downwards by means of the reflector in which the oscillator is mounted (Figure 39). The reflector itself is made from two thin conical metal sheets soldered together

THE USE OF SOUND WAVES

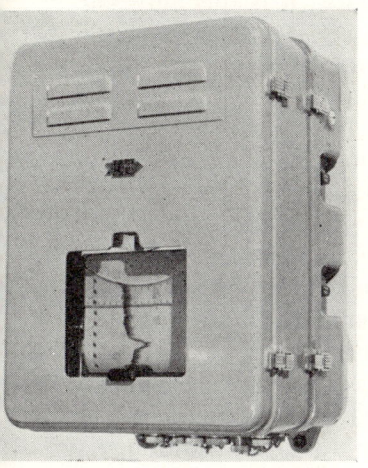
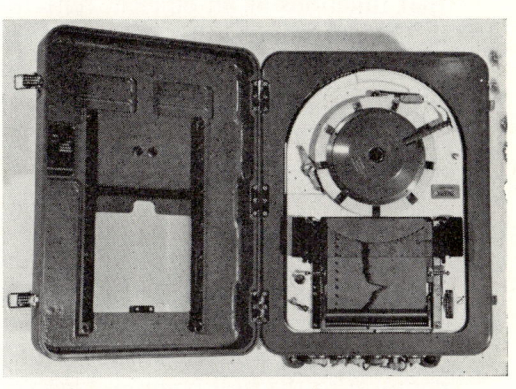

Figure 38

Echo sounder Kelvin & Hughes Type M.S. 26 Survey Echo Sounder open and shut

(*Courtesy of Messrs Kelvin & Hughes*)

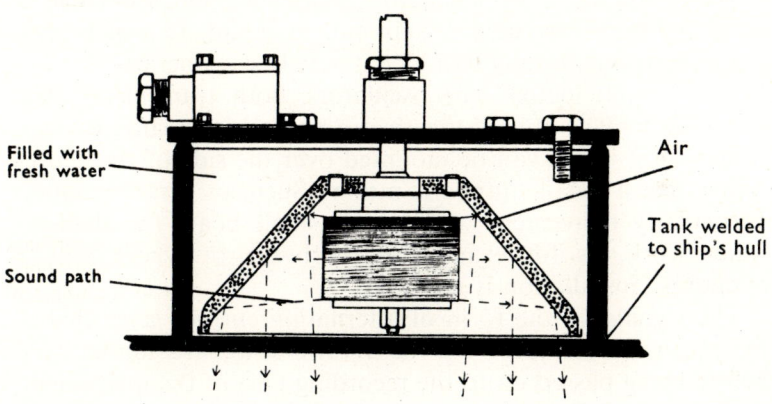

Figure 39

A magneto-striction oscillator. The nickel laminae mounted inside the reflector which consists of liquid or air between thin metal plates and enclosing between them an air space; it is this air space that functions as an efficient reflector.

The receiving oscillator (when a separate unit) is, so to speak, a transmitter in reverse and indeed looks identical, but the nickel stampings of the former have their small amount of residual magnetism increased by 'flashing' the windings. However, in some modern installations the transmitting and receiving elements are combined into a single so-called transceiver. The return echo impinging on the receiving oscillator causes vibration in the nickel stampings and these induce in the winding an alternating voltage which may be amplified and passed to the recording part of the instrument.

In steel ships both transmitter and receiver tanks are cut to conform to the hull being welded round their circumference to the plates, with the reflectors taking up a vertical position. As already mentioned, no holes need usually be cut in the ship's plates; the intensity and wavelength of sound in steel are such that, at ultrasonic frequencies, sound waves up to about 15 kc/s will pass through the normal thickness of ship's plate without any serious attenuation, provided there is a good sound-conducting layer of water on both sides of it. (At high frequencies pierced-hull oscillators may be obligatory.) The oscillator tanks are always kept full of liquid, to which anti-freeze components may be added if very low temperature conditions are anticipated. The oscillators, both transmitting and receiving, may be housed together in a separate streamlined case (Figure 40), which can be mounted over the side of the ship—rather like a small outboard motor. Such an arrangement is particularly convenient when using small boats for shallow-water work, as for example in harbours or where only a temporary installation is required.

The signals, in the form of alternating current generated in the receiving oscillator by the return echo, are minute and before being passed on to the recording part of the instrument, must be amplified. In practice, the amplifier is often a self-contained unit fitting in, or mounted close to, the side of the recorder unit. In many respects the amplifier circuit follows normal radio practice except that it is designed to give maximum

THE USE OF SOUND WAVES

amplification at a fixed frequency. Two stages of amplification, partially under manual control, are usually included; the total gain at maximum sensitivity is about a million to one. A rectifier may be incorporated in the amplifier unit to convert the amplified alternating into direct current, in which form it is required for some types of recording paper.

So far we have seen how the ultrasonic impulses are generated, transmitted, and received, and how the return signal is amplified. We must now consider in more detail the recording. As already noted, the speed of sound in water is about 4,900 ft. per second, so that a time interval of one second between transmission of the sound wave and reception of the echo corresponds to a depth of about 2,400 ft., or 400 fathoms. A modern instrument is capable of measuring fractions of a fathom, and it incorporates, therefore, a precision chronometer.

Figure 40
Outboard unit for echo sounding
(Kelvin & Hughes)
(*Courtesy of Messrs Kelvin & Hughes*)

There are several ways of recording the return signal in relation to depth, but all employ the same principles. In general, a moving pointer travels over recording paper. As the pointer passes the zero mark, a switch is operated and a pulse of sound transmitted. By the time the return signal is received and passed to the pointer the latter will have moved a certain distance, which if it moves at a constant speed, will be proportional to the time taken for the echo to return and hence to the depth. A mark is made on the paper by the return signal. By placing a calibrated scale against the pointer the depth can be read off directly. In some instruments the moving pointer or stylus is mounted on an arm which travels in a circle, in others it is mounted on an endless band which moves across the paper (Figure 41).

The range and calibration of the depth scale is directly related to the speed of movement of the stylus. For example, if it takes exactly one second for the arm to cover the full length of a given scale, then that scale might be divided into 400 divisions, each of which represents one fathom. Clearly, by suitable selection of stylus speed, any required scale range is possible and it may be calibrated in feet, fathoms or metres. Also, by providing in one recorder a choice of two alternative gear ratios for the movement of the arm, one of which is, for example, six times the other, a single scale may be utilized to indicate feet at the higher speed and fathoms at the lower.

In the early instruments—a few are still in use—the depth is indicated on a circular scale evenly divided into fathom intervals. A gas-filled relay tube rotates on the arm and when the signal is returned the scale is lit up at that point by a brilliant flash.

In most modern instruments recording paper is drawn through the instrument and successive echo marks appear side by side and build up into a thick line, giving a continuous record of the depth of water and following every undulation of the sea or river bed (see Figures 43-46). The mark may be made on a moist recording paper treated with potassium iodide and starch; the current transmitted to the stylus electrolyses the iodide, producing a lasting brown stain of a complex starch iodide. Alternately, a coated graphite paper may be used and this is burnt at the stylus point to give a black trace.

Cathode ray tube presentation

In some recent instruments the return echoes are presented on a cathode ray tube which may replace or be additional to the recording paper. In one setting a large depth, 100, 200 or 500 metres, may be recorded, but by turning a switch a small depth

THE USE OF SOUND WAVES

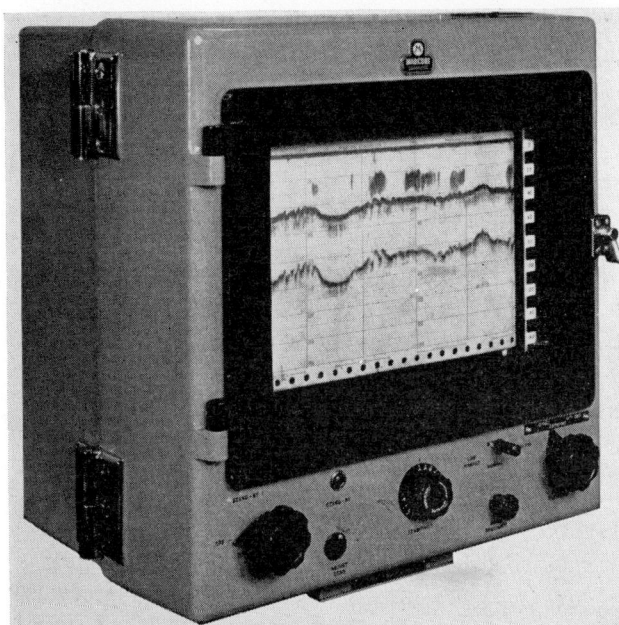

Figure 41

Interior and exterior views of Marconi echo sounder. In this machine the writing stylus moves in a continuous linear band across the paper. Note fish traces and double echo

(Courtesy of the Marconi International Marine Communication Company Ltd)

interval may be selected for detailed observation. This method has some advantages, particularly in biological work, since the intensity of the echo is related to the magnitude of the return 'trace' (see below). Further, it facilitates the detection of traces from fish near the bottom even though the latter is rough. Unless some additional recording camera unit is attached, as may be done for research work, the cathode ray tube does not, however, give a permanent record.

The cathode ray tube consists of an evacuated bulb in which is mounted an electron gun (Figure 42), and two sets of deflecting plates, one pair being mounted horizontally and the

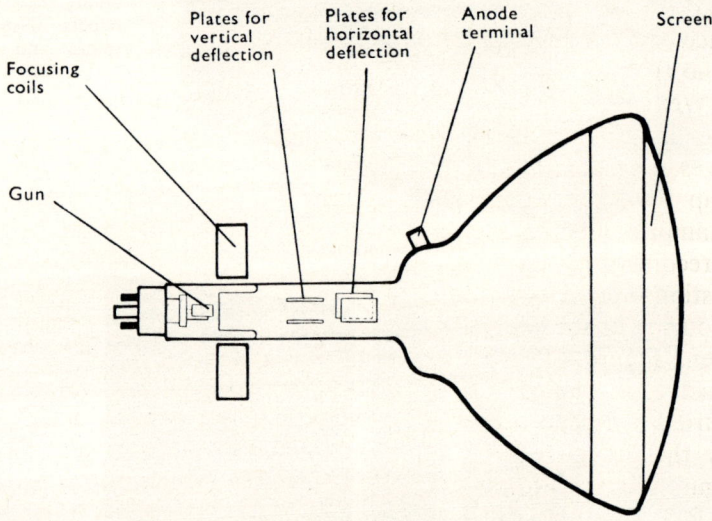

Figure 42

Cathode ray tube

In the cathode ray presentation of echo sounding the return signal is applied to the plates for horizontal deflection

other vertically. The cathode gun gives a stream of electrons. When these impinge on the face of the tube, which is coated with a fluorescent material, they cause it to glow; where the electron beam impinges, a bright spot is seen. When a potential is applied to the plates, the electron beam is deflected and the

THE USE OF SOUND WAVES

spot will move up, down or across the screen according to which plate the potential is applied. Instead of the transmission pulse being sent out when a stylus passes zero—as in the recording instruments just described—a deflecting potential is applied to the horizontal plates (Figure 42). In the absence of an echo a vertical bright line is, therefore, seen on the face of cathode ray tube. The echo after reception and amplification is fed to the vertical plate and a horizontal deflection of the spot thereby effected. The vertical movement of the spot is the time base and this may be selected for various depth intervals. In the absence of any object in mid-water the bottom appears as a series of horizontal lines, the depth being read on a scale in front of the cathode ray tube. The appearance of fish is indicated by horizontal lines above the bottom echo (Figure 43).

We have followed the development of the echo sounder from the 'knock and listen' stage to the modern machine which gives a continuous record of the depth of water beneath a ship under way, and this section may be closed by a few examples of some of the more striking results. Figure 44 shows a record of the *Lusitania*. The wreck, some 80 ft. high, is seen resting on a flat bottom in just over 300 ft. of water and the scouring of the sea bed to one side of the wreck can be clearly seen. Figure 45 gives some idea of the detail that may be resolved in shallow waters; it shows a record of a survey taken during a dredge operation, and the amount of mud removed by the dredge can be quite accurately estimated. The third figure (Figure 46) shows a section in the far south taken on R.R.S. *Discovery* during her work in the Antarctic. It indicates a varied bottom between 750 and 2,040 ft. in the Schollaert Channel of the Palmer Archipelago, only a few miles from the edge of the Antarctic Continent.

THE ECHO SOUNDER AND FISHERIES

So far we have considered the echo sounder only as a depth-determining device; we may now consider its use in the fishing industry and fishery research.

OCEANOGRAPHY AND MARINE BIOLOGY

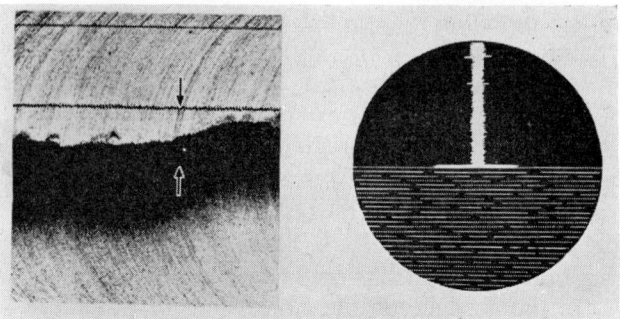

a) A well-defined single echo right on the bottom is clearly shown on the tube face, although the record-chart shows no discernible trace

b) A fairly heavy collection of fish in mid-water is indicated by the conglomerate echo-type signals shown here. The full extent of the shoal is shown by the shaded area on the record chart

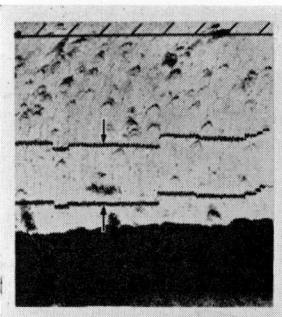 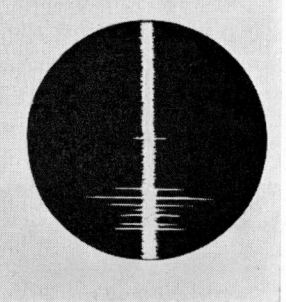

c) A multiple echo-type signal is shown on the tube face, indicating the presence of a small fish shoal in mid-water. An equivalent shaded patch is visible on the record chart

Figure 43

Echo-graphs and cathode ray tube records

(*Courtesy of Messrs Kelvin & Hughes*)

THE USE OF SOUND WAVES

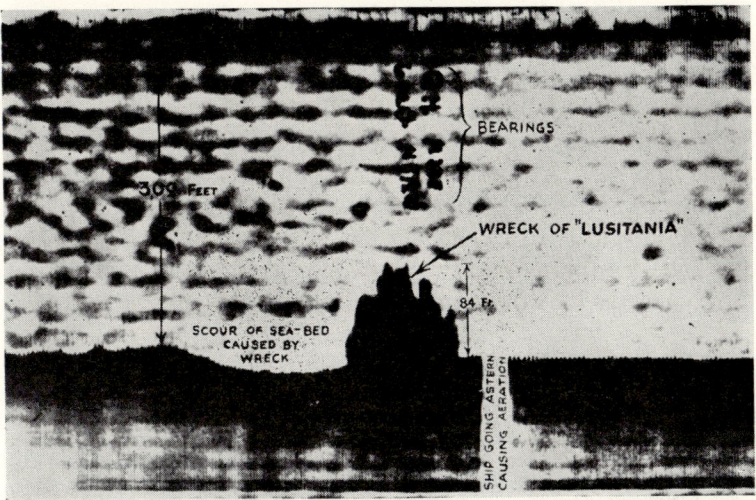

Figure 44
Wreck of the *Lusitania*
(*Courtesy of Messrs Kelvin & Hughes*)

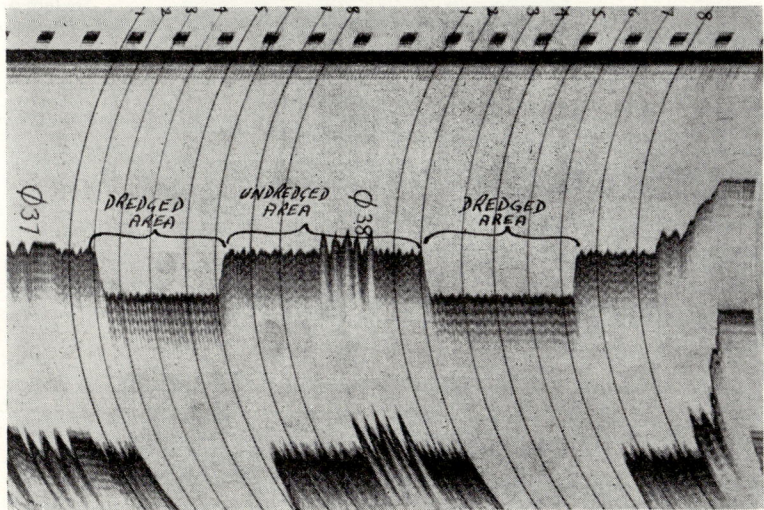

Figure 45
Tracing in a channel showing the material removed from the bottom during dredging. The volume and depth cleared can be determined. Note the double echo. Time marks are at the top
(*Courtesy of Messrs Kelvin & Hughes*)

OCEANOGRAPHY AND MARINE BIOLOGY

It is perhaps not generally realized that, quite apart from any navigational value, to know the depth of water is of great importance to a fishing skipper—particularly a trawler-man. Fish taken by trawl are usually over more or less shallow banks surrounded by deeper water; these banks have first to be found and then, while the trawl is being worked, the ship must be kept on the bank. This is necessary not only to catch fish, but to prevent damage to the nets as a result of their 'running off' into deep water and then having to be dragged back across the rocky edges of the bank. Nets are very expensive and it is

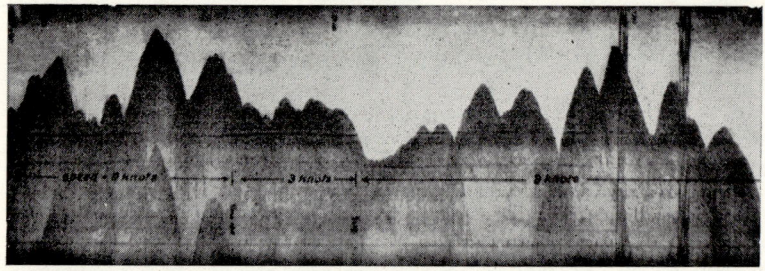

Figure 46
Echo traverse in Antarctic, Schollaert Channel, Palmer Archipelago
(*From* H. F. P. Herdman, '*Discovery Reports*', 1953. *Courtesy of The National Institute of Oceanography*)

therefore not surprising that even in 1928—early days for the echo sounder—an instrument using an electric hammer and hydrophones was installed in a Hull trawler. It was not, however, until ultrasonic sounding had been developed that the machines came into general use on commercial fishing vessels; by 1935 some hundreds had already been installed in British ships alone. The development of continuous recording instruments, giving a permanent record, was an important advance. With their introduction the skipper had no need to rely on the mental picture of his fishing grounds built up over many successive trips; he had available a continuous record that could be consulted whilst working the ground or scrutinized more carefully at leisure. Initially echo sounders were installed only on trawlers, where they were still essentially used as a navigational aid both

THE USE OF SOUND WAVES

in locating the fishing banks and, once over them, in manipulating the trawl. It is certain, however, that in this way fishermen had become echo sounder conscious and were the more ready to appreciate any new potentialities of the instrument.

Echo traces from fish

A number of people had from time to time suggested that fish could give rise to echoes and echo traces, and it was in 1933 that the Ministry of Agriculture and Fisheries' Lowestoft research ship *Onaway*, using a flashing-type Marconi echo meter, noticed distinct 'pulse echoes', as they were called, occurring constantly at one depth between surface and bottom; the suggestion was again made that they were caused by fish. Two years later, the Norwegian fisheries scientist, Oscar Sund, using one of the newer recording machines, reported that he had detected spawning cod at Høla in the Lofoten area, and for a long time it was thought that he was the first to obtain recordings of fish on echo sounder paper. It has since been found that another Norwegian, Skipper Bokn, published traces some eleven months before Sund, showing sprat shoals at the surface in the Frafjord. The Lowestoft workers discussed their results and suspicions with a herring fishing skipper, Mr Ronald Balls. He had been closely connected with fishing from childhood, for his father was a trawler owner and he himself had inherited an interest in his father's business. He was a practical fisherman whose job it was to catch fish for a living. As a result of these discussions, Skipper Balls began to study carefully the flashes from his echo meter while working at the herring fishing; by the end of his 1933 season he was convinced that the echo sounder was capable of recording the presence of herring shoals and that it could give the depth at which the shoal was moving as well as some idea of its density and extent. This was something quite new.

These initial trials and subsequent intelligent observations lead to a much more detailed survey of the possibilities of the instrument in herring fishing. Skipper Balls kept detailed records of echoes and catches during his drift-net fishing over the four years 1933-37. During the summer period his ship shot 69 times 'on echoes' and the average catch was 26 crans of herring; this was more than three times the average catch of 8 crans that he obtained when shooting without an echo. In giving an account of these results he was careful to point out that shooting on an echo did not *guarantee* a successful catch; there are many other factors to be taken into account besides the production of an echo—not least the behaviour of herring themselves. Nevertheless, the gain achieved over a season by echo fishing was quite clear; it has since become standard practice in herring fisheries. We may quote an account written by Skipper Balls himself of one night's fishing: 'Once, at 147 miles NNE. of Shields herring were echoed almost continuously whilst steaming over a fairly large area at 6.0 p.m. It was a dull southerly kind of evening. The spotline [echo sounder] showed many separate shoals, sometimes at surface, mid water and bottom simultaneously, but no particular echo lasted more than 10 or 12 soundings. Shooting here brought 150 crans. This was the only case of such numerous echoes and was the biggest shot hauled in the period of summer fishings.'

The echo sounder saves time; herring, for example, are located more quickly and less time is lost casting where there are no fish or where the shoals are small. The great majority of all fishing vessels now carry echo sounders and there is no doubt that they have helped to increase the efficiency of both the herring and cod fisheries. In their cod fishery in the Lofoten area, following requests to the authorities by the Norwegian fishermen themselves, at least one vessel is employed solely to locate fish and to radio information, by reference to special grid charts, to the fleet.

The relation of echo trace and fish species

It is, of course, very desirable to know what kind of fish are giving rise to any particular echo trace and much research has been and still is being done on this subject. In Britain, scientists of the Ministry of Agriculture, Fisheries and Food, at Lowestoft and those of the Scottish Home Department at Aberdeen have been particularly active in this field, and comparable work is

THE USE OF SOUND WAVES

reported from most nations with an interest in fishery biology. Expert observers can now often make more than a good guess as to the type, if not the actual species, of fish giving rise to a particular echo trace in any specified area. In 1950 Dr W. C. Hodgson of the Lowestoft Laboratory published the first account of efforts to relate the different kinds of traces to the particular species of fish responsible. In many cases he identified the fish by capture and in this way traces characteristic of pelagic fish, such as mackerel, herring, pilchard and sprat, as well as of the demersal gadoid fishes, cod, coalfish and pollack, were determined with some certainty. Some of Dr Hodgson's traces are reproduced in Figures 47 and 48.

Figure 47
Herring shoal as recorded on the 'Seagraph' sounder in the M.V. *Platessa*
(*From W. C. Hodgson, 'Fishery Investigations', Series II. Courtesy of the Ministry of Agriculture, Fisheries and Food, Lowestoft*)

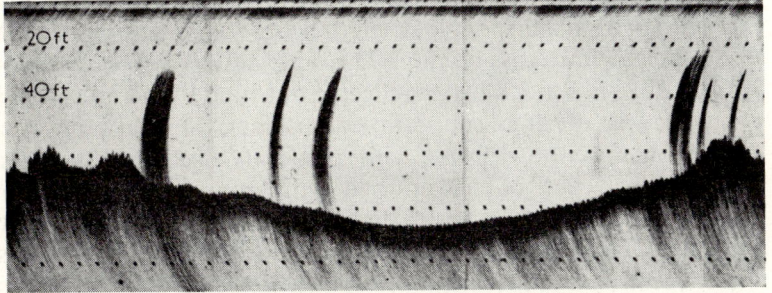

Figure 48
Series of isolated pilchard shoals off the Cornish coast
(*From W. C. Hodgson, 'Fisheries Investigations', Series II. Courtesy of the Ministry of Agriculture, Fisheries and Food, Lowestoft*)

There was little doubt about the species producing the traces shown in the Figures, for fish were actually caught and identified: in many other cases 'identifiable' marks are known either from catches taken at the time or, with less certainty, from general information regarding the fish in the area.

The problem of correlating trace and fish is, however, proving rather more difficult than was anticipated when the work was begun. The production of a trace is a very complex process and recent investigations have shown that the same fish can give rise to different kinds of traces according to the conditions and the instruments used. We have to consider the instrumentation in relation to the target. The most important considerations regarding the former are speed of the recording paper, speed of the ship, frequency of the sound, and beam angle. Even with a given shoal of fish at a given depth, any change of these variables will affect the type of trace. It is evident that paper speed and ship's speed will alter the trace since they determine when the marks are made on the paper. The beam angle is important in relation to the size of shoal as distinct from its density. The frequency of the sound waves must be considered in relation to the size of the fish; when its cross-section approximates to the wavelength used a stronger signal is obtained.

With regard to the fish, their dimensions, character, shoaling density, size of shoal and depth below the ship have all to be taken into account. A large fish produces a stronger return signal than a small fish—the strength being roughly proportional to its horizontal cross-section. The air bladder—since a water-air interface is a much better reflector of sound than a fish-water interface—is of great importance: in fact, Dr Cushing has shown that in those fishes with moderate-sized air bladders about 40 per cent. of the returned signal may be attributed to the air-water interface, even though the air bladder takes up only 5 per cent. of the total volume of the fish. Such estimates were confirmed experimentally using artificial air bladders. These results explain why mackerel, which do not possess an air bladder, give such a thin trace. The number of fish per unit volume, that is, shoaling density, and their depth below the ship

THE USE OF SOUND WAVES

must be taken into account: the strength of the signal is roughly proportional to the number of fish on the axis of the sound beam. One of the main differences between herring and pilchard traces, for example, is often the intensity, and this may be a consequence of the tighter shoaling of pilchards.

Application to research work

The echo sounder has also been used in studying the behaviour of fish; for example, vertical migration of fish was shown by Dr Runnström in 1941 and by Dr Tester in 1943. Mr Richardson of the Lowestoft Laboratory has followed the behaviour of fish over quite long periods. He was able on one occasion to keep the ship over the same shoal of sprats for 24 hours while running the echo sounder continuously. The echo trace clearly showed that the shoal rose after dark until it was up in the surface layers (when the noise of the fish breaking the water could be heard), but with increasing light at dawn the fish descended. He could not follow a *single* shoal of herring, but an examination of echo traces taken over a period during the North Shields fishery showed that herring too were more often in the lower layers during the day than during the hours of darkness.

In addition to making extended observations followed by attempts to interpret and correlate the results—a common form of attack in many biological problems—the echo sounder can be used to record the behaviour of fish in experimental work: the environment is artificially modified and the effect followed by means of the echo trace. For example, the effect on fish of lights of various colours and intensities, and of different kinds of shock, have been investigated, the result of the stimulus being followed in detail by the echo sounder. Figures 49 and 50, taken from Mr Richardson's work, show that after switching on a light at night a shoal gradually collected: after the light was switched off, the fish began to disperse. It is also clear that the first reaction to light of the residual shoal was a 'shock' reaction —a slight descent into the water—followed by a rise and then aggregation.

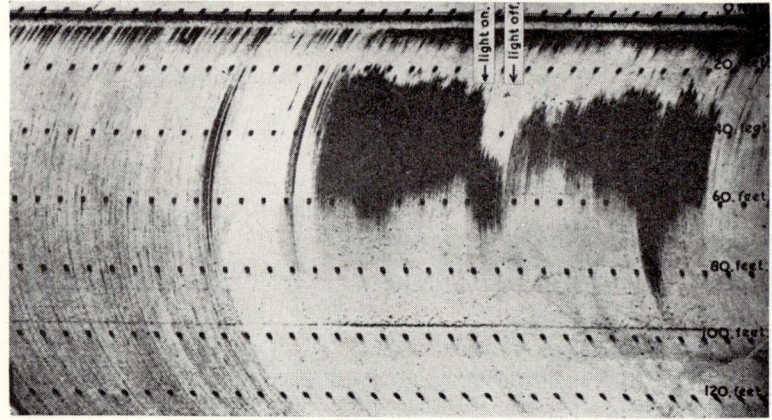

Figure 49

The effect of a searchlight shone vertically on a shoal of herring in East Anglian waters is shown in this tracing; note that shoal appears to move bodily down when light switched on

(*From I. D. Richardson, 'Fishery Investigations', Series II. Courtesy of the Ministry of Agriculture, Fisheries and Food, Lowestoft*)

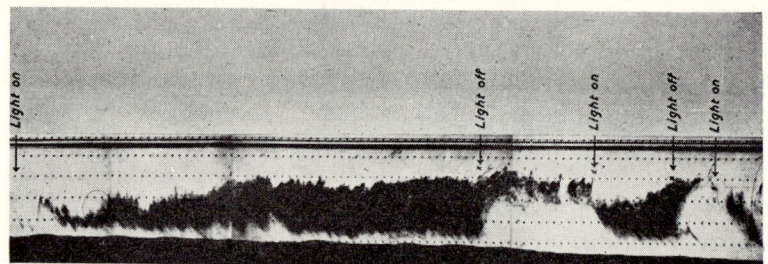

Figure 50

A tracing obtained off Cornwall in which a shoal of fish is collected by the searchlight. The points at which the lights were switched on or off have been marked on the chart

(*From I. D. Richardson, 'Fishery Investigations', Series II. Courtesy of the Ministry of Agriculture, Fisheries and Food, Lowestoft*)

Deep scattering layer

All the traces we have just been discussing are known with a fair degree of certainty to originate from fish. There is, however, another type of trace, first noticed by a war-time underwater sound group of the U.S. Navy, and this has aroused great interest. Records were first obtained over very deep water when the trace itself was often as much as 50 fathoms thick; it was frequently continuous over wide areas—so much so that it presented the appearance of a false bottom. It was called the Deep Scattering Layer. Further work has shown that there may be a number of such deep-water traces whose behaviour is not always the same. These scattering layers have now been recorded by many workers, not only from the deep-water regions of all the major oceans but also from shallow water. Apart from continuity, one of their most interesting features is that they frequently ascend at sunset, stay in the upper layers (usually with some dispersion) during the night and then descend during daytime.

The universal appearance and the wide extent of this type of trace might suggest that it was caused by some sharp physical change such as density but, on the basis of any known physical phenomenon, it is extremely difficult to account for the frequently observed night ascent and day descent. However, many animals are known to perform such daily vertical migrations, and it seems almost certain that in general the trace is of biological origin. Which organisms are responsible is much less certain. Dr H. B. Moore has presented evidence that in deep oceanic regions planktonic organisms, such as the shrimp-like Euphausids, are responsible for this deep scattering layer. However, certain physical parameters are necessary to account for the observed scattering and it is not yet certain that Euphausids have the required properties. It may even be due to deep-water fish feeding on these smaller planktonic organisms and moving up and down with them, in which case the air bladders of the fish may be the primary cause of the scattering. There is good evidence that small fish are responsible for the scattering layers observed in shallower waters. The most recent work strengthens earlier suggestions that the layer is not always caused by the same organisms—that, in fact, there exist different types of these deep scattering layers in different places and at different times. It is a fascinating problem and one which might be readily solved if only all the appropriate techniques could be brought together in the right place at the right time.

Asdic and fish detection

All the above work has related to the use of a sound beam projected downwards. In Asdic the echo sounding principle is used for detection in a horizontal direction; the oscillators can be rotated and the impulses transmitted horizontally in any given direction. In this way shoals of fish should be detectable along a bearing line up to some 2,500 yards from the ship. The Norwegian vessel, *G. O. Sars*, which was commissioned in 1950, carries Asdic, and using this instrument, migrating herring have been followed for distances up to 300 miles. On one occasion fishermen caught herring detected in this way and valued at seven million kröner; the 'normal' fishing grounds were giving poor catches.

Echo fishing is still relatively new but, to quote from a recent F.A.O. Fisheries Bulletin issued by the United Nations, 'it is safe to say that of all the tools of this trade acquired during this period (since World War II) the common use of the echo sounder for detecting fish must be considered the outstanding development of commercial fisheries'.

UNDERWATER NOISE

The sea has often been referred to as a silent world—but read Mr Fisher, recounting in 1819 his experiences off the Prince Edward Isles in the Canadian Arctic and recorded in his *Journal of a Voyage of Discovery to the Arctic Regions*: 'Whilst we were pursuing them [white whales] to-day, I noticed a circumstance that appeared to me rather extraordinary at the time, and which I have not indeed been able to account for yet to my satisfaction. The thing alluded to is a sort of whistling noise that these fish

THE USE OF SOUND WAVES

made when under the surface of the water; it was very audible, and the only sound I could compare it to, is that produced by passing a wet finger round the edge or rim of a glass tumbler. It was most distinctly heard when they were coming towards the surface of the water, that is, about half a minute before they appeared, and immediately they got their head above the water the noise ceased. The men were so highly amused by it, that they repeatedly urged one another to pull smartly, in order to get near the place where the fish were supposed to be, for the purpose of hearing what they called a "whale song": it certainly had very little resemblance to a song, but sailors are not generally the most happy in their comparisons.' This is an extract from Mr Fisher's account of the voyage of His Majesty's Ships *Hecla* and *Griper*, sent out to discover the North-West Passage. Nor is this the only reference by early explorers to sea noises; Dr Kane, surgeon on board the ship sent out to search for traces of Franklin's expedition, writes in 1854: 'On this occasion I heard the white whale singing under the water—a peculiar note between a whistle and a Tyrolean yodel. Our men compared it to a Jews-harp.' Little wonder that the white whale or dolphin—Begula, from the Russian signifying white—became known as the sea canary. A silent sea—not at all! Even the very names of some fish—croaker, drum-fish—suggest their vocal character, and the fisherman from Malaya who puts his head over the side of his prau, holding it under the water and waiting for fish 'honks' before shooting his nets, has, like the ancient Phoenician fishermen who located shoals of drum-fish by their sounds, little doubt about the reality of underwater noise.

However, marine biologists took little interest in the occurrence of natural sea noises until their widespread occurrence and importance was vividly brought to the notice of naval scientists early in World War II. Soon after its outbreak an extensive and intensive programme of underwater listening was initiated in connection with the detection of underwater craft. The apparatus consists of a set of submerged microphones, i.e. hydrophones and an ordinary receiver similar in principle to that used in a dictaphone. The sounds may be recorded on tape or discs. Frequency analyses are made by a harmonic wave-analyser. We will follow the adventures of two U.S. Navy Groups—one on the east coast and the other on the west—in both cases charged with maintaining sonic stations at the entrance to important harbours. They soon found, as indeed would be expected, that various human activities in harbours—ships' engines, pile driving and the like—give rise to underwater noise. They also quickly realized that in the open oceans, wave motion—particularly the number and prominence of breaking waves and white caps—was responsible for much of the so-called ambient or background noise which could, therefore, be directly related to weather conditions; even rain was found to give rise to underwater noise and some recorded sound was ascribed to the rolling of pebbles and gravel. Quite dramatically in 1942, very strong 'interfering' noises became troublesome in the underwater sound-detecting system of Chesapeake Bay. Here, a number of regularly spaced hydrophones had been connected to a shore station so that not only could long periods of listening be undertaken but signals could be continuously recorded on instruments.

A 'single' source of this unusual interference often gave a drumming sound, but the overall effect resembled that of a pneumatic drill tearing up a pavement; there was no doubt that it was quite different from any noise previously detected from adjacent sources. Detailed investigation of the records and continuous listening over long periods showed that there was a diurnal rhythm in the intensity of this 'interference'—the noise level rose at night and fell during the day. It seemed possible that the noise was of biological origin and a small charge was therefore exploded under the water; immediately the noise ceased—to begin again quite suddenly. At about the same time fishermen reported large quantities of croaker fish in the vicinity—one estimate gave 300 million fish in the Bay. Some of these fish were captured, transferred to an aquarium and tested for their vocal characters. A comparison of these results with those obtained from the open Bay quickly confirmed the hypothesis that the Chesapeake Bay 'chorus' did indeed originate from dense shoals of croakers.

THE USE OF SOUND WAVES

The amount of noise from these croakers was about half as much again as that normally found in the vicinity. The distribution of the noise over a whole range of frequencies was then investigated and, as can be seen from Figure 51, most of it was confined to frequencies below 2,500 cycles per second—quite

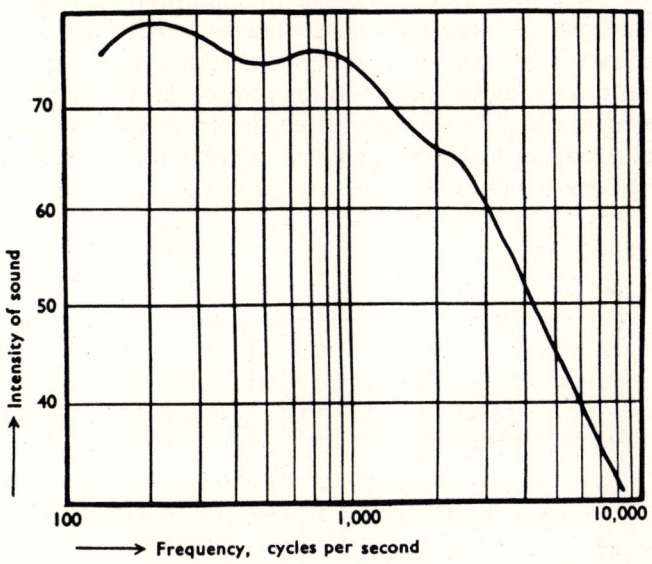

Figure 51
Frequency character of croaker noise
(D. P. Loye and D. A. Proudfoot, 'Journal of the Acoustical Society of America', 1946; Redrawn from original)

distinct from the frequency distribution of noise from a calm sea. A frequency analysis of croaker noise made later in the same season showed that much of the high-frequency noise associated with the fish had disappeared—the noise had become concentrated towards the lower-frequency bands. The known growth rate of the fish is not sufficiently high to indicate that development was rapid enough for increased size of the fish to cover such a change; a more likely explanation is that a population of older fish had migrated into the area, replacing the younger generation.

OCEANOGRAPHY AND MARINE BIOLOGY

No sooner had the mysterious noise in Chesapeake Bay been laid at the door of the croakers than a listening group on the Pacific coast were faced with the same problem—a disturbing amount of interfering noise in their hydrophones. Perhaps with the experience of the Chesapeake Bay group in mind, or perhaps because there were more biologists to hand, a biological origin was immediately suggested and a whole series of animals from the vicinity were collected and put through aquarium tests. The noise was quickly traced to the snapping of the claws of a group of non-edible shrimps. Over the hydrophones a single animal produces a sharp 'crack' or 'snap' while the combined effect of large numbers gives a continuous crackling sound.

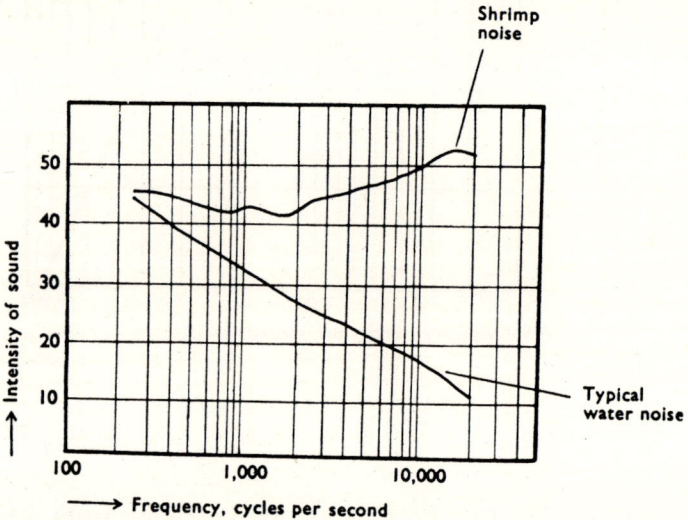

Figure 52
Frequency character of snapping shrimp noise compared with that of typical ambient water noise spectrum in quiet situation
(D. P. Loye and D. A. Proudfoot, 'Journal of the Acoustical Society of America', 1946; Redrawn from original)

The noise spectrum was again investigated (Figure 52) and it may be compared with that of the croaker noise in Figure 51. Clearly the shrimp noise is concentrated far more in the higher frequencies than the croaker noise.

THE USE OF SOUND WAVES

The origin of this second underwater sound has been investigated in some detail. The claw structure in all species of the shrimp genera concerned (*Crangon* and *Synalpheus*) is similar, and to some extent they are probably all capable of making a snapping sound. This snapping habit is associated with both defensive and offensive activities since, when the claw snaps, a strong jet of water is shot out; this probably frightens away other animals. The two claws of the shrimp are often quite different in construction, the sound-producing snapping claw being much the larger; indeed, in some species it is nearly as large as the whole body of the animal, so that there is ample

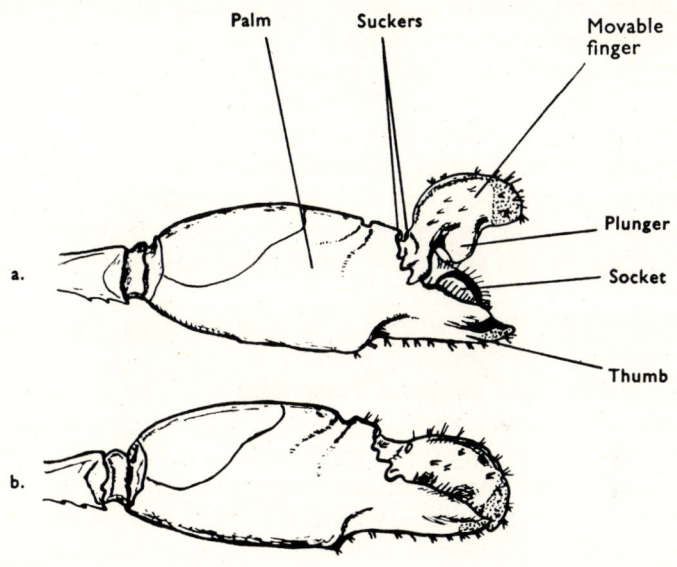

Figure 53

    Snapping claw of shrimp *Crangon californiensis*
      a) Snapper open   *b*) Snapper closed

room for development of strong snapping muscles. The detailed structure is shown in Figure 53. Near the outer end of the 'palm' is a movable hard 'finger' with an immovable 'thumb', forming a pincer-like arrangement. On the inner side of this finger is a 'plunger' structure fitting into a socket on the thumb, and

99

leading from this socket is a groove through which water is expelled when the plunger is suddenly forced into the hollow.

This plunger-socket device is present in both *Crangon* and *Synalpheus*. All species of *Crangon* have in addition a so-called 'sucking' disc on the outer side of the finger near its point of attachment (Figure 53). When the finger is fully raised and ready to snap, the disc contacts on a precisely similar one on the palm, the two being held firmly together. It is believed that these suckers serve as a trigger holding the finger back, and that extra muscular tension has, therefore, to be exerted to break the contact, with a consequent increase in the force of impact. Such suckers are absent in the genus *Synalpheus*, but in the same relative position and probably serving the same purpose there are smooth contacting surfaces.

As the finger is raised the plunger is withdrawn from the socket, and it has been held by some observers that the noise is produced during this withdrawal—rather like the familiar 'plop' on drawing the cork from a bottle. The function of the plunger-socket arrangement does not appear, however, to be the production of sound but rather the ejection of a stream of water as already described. (It may also aid in preventing dislocation of the finger when it is violently snapped.) The precise way in which the sound is produced still appears a matter of some doubt, but the most reasonable explanation is that it results from impact when the movable finger strikes a glancing blow on the opposing tip of the thumb. It is of passing interest to note that the snapping claw may be either the right or the left; if lost through injury, the inconvenience is only temporary and a new one is always grown in subsequent moults. It is a curious fact, however, that in the renewal process the lost large snapping claw becomes replaced by a small pinching claw, while the original pinching claw is enlarged to give rise to the new snapping claw.

We have concentrated on two sub-surface noises of animal origin largely because of the sudden and somewhat dramatic way in which they were encountered. Sound production is now known to be of common occurrence in marine fishes and much work has been done on this subject in the past few years. The organ most commonly concerned is the air bladder, a membranous sac lying in the body cavity between the backbone and gut. Its form is diverse and the type of sound produced is particularly a function of the size and shape of this organ, which may be vibrated in a number of ways. Air-bladder sounds are usually low pitched, guttural, vibrant, and drum-like, and exhibit a frequency range from 50 to about 1,400 cycles per second, although most of the sound energy is concentrated towards the lower part of the spectrum. Friction of some part of the body, most frequently the 'gill teeth' within the throat or the jaw teeth, gives rise to stridulent sounds of a rasping, scratching, scraping or whining character. Again, although the frequency range may be very wide, from 50 to 4,800 cycles per second, the maximum energy with this type of sound is concentrated higher in the spectrum than is the case with air-bladder noise. In the noisiest fish, both organs take part in sound production. The biological function of these sounds is uncertain; they may bring individuals of a species together for mating and spawning or colonial activities; they may be used as a defence mechanism; it has even been suggested that in some deep-water fishes they may act as an echo sounder and enable the fish to determine its depth above the sea bed. It is certain that many of these fish are well provided with sound receptors and that these are particularly sensitive to the low frequencies which they themselves emit.

An interesting field of work has been opened up by these investigations in which co-operation between biologists and physicists is essential and most profitable. We are particularly concerned with the possibility of using underwater sound-detecting equipment as a biological tool. It is clear that if the noises are specific to certain animal groups, then underwater listening may be used to plot the distribution of such animals. It has even been suggested that hydrophone equipment may be used by fishermen rather in the manner of an echo sounder. It must be pointed out, however, that although many fish have been shown to emit noises under laboratory conditions (and on removal from water) and whilst it is not unreasonable to assume that they will always emit similar noises with the appropriate stimulus, only a limited number of noises have, in fact, been obtained under entirely natural open-sea conditions. Further, the noises even from known sources are not continuous; from the fisherman's point of view this is a distinct disadvantage. Again, propeller noises constitute a severe source of interference and the ship would, therefore, have to be stopped during the observations. Nevertheless, as a tool for research purposes hydrophone techniques have distinct possibilities and some of these may be illustrated by reference to the work already done on the snapping shrimp itself. Figure 54 shows a plot of the noise intensities obtained in hydrophone traverses across a stretch of variable inshore bottom near San Diego. It is clear that the amount of noise is directly related to the type of substratum being traversed. Over the sand and mud there is little shrimp noise but over the zone of shale, which is honeycombed by abandoned mollusc borings and over cobbly or boulder-covered ground the noise is considerable. The intensity of sound over the rough ground clearly indicates the presence there of a large number of snappers. By contrast, dredging produces very few shrimps and indeed gives the impression of quite a small population. Now, the animals are relatively shy creatures and although capable of swimming rarely do so in the adult stage; rather, they seek shelter and concealment in crevices between stones and boulders or in holes in the shale. Collecting by dredges over such rough ground is notably ineffective and it is clear that the discrepancy between the sound and dredging techniques is really due to the inadequacy of the latter over such ground; this in turn is dependent upon the habit of the shrimp living in such places. There is no doubt that a more accurate measure of the distribution of the shrimp population is obtained by the listening technique than by dredging. The intensity of sound is related to shrimp numbers and the method may be used to plot the distribution of the animal in other areas. It has been suggested that shrimp noise could be used to locate sponge beds—in which some shrimps live—prior to commercial exploitation. Deep-water forms of these shrimps are known, but it has been found that the noise intensity even over suitable rocky ground, where they might be expected to occur, decreases rapidly below 30 fathoms

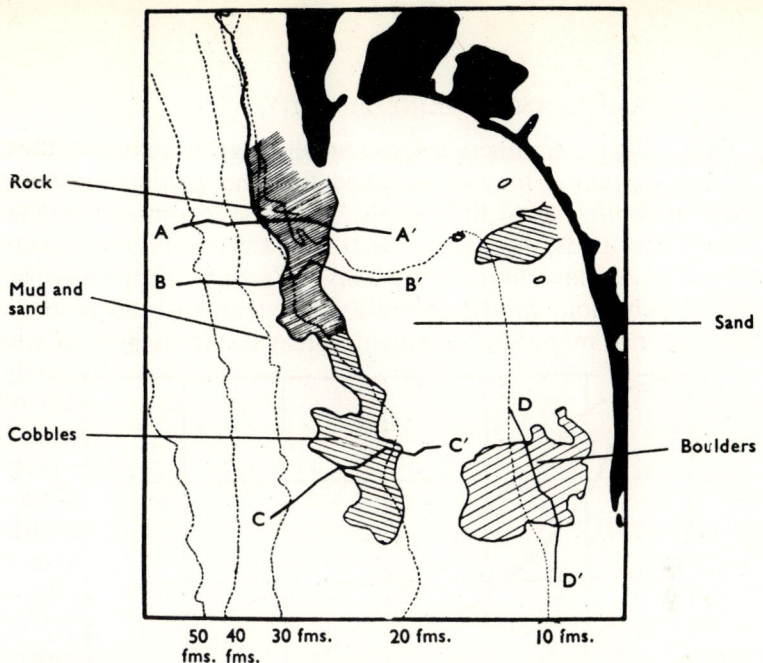

Figure 54
Change in noise level over a series of traverses. Different types of bottom are indicated by the same shading

(M. W. Johnson, F. A. Everest and R. W. Young, 'Biological Bulletin', 1947; Redrawn from original)

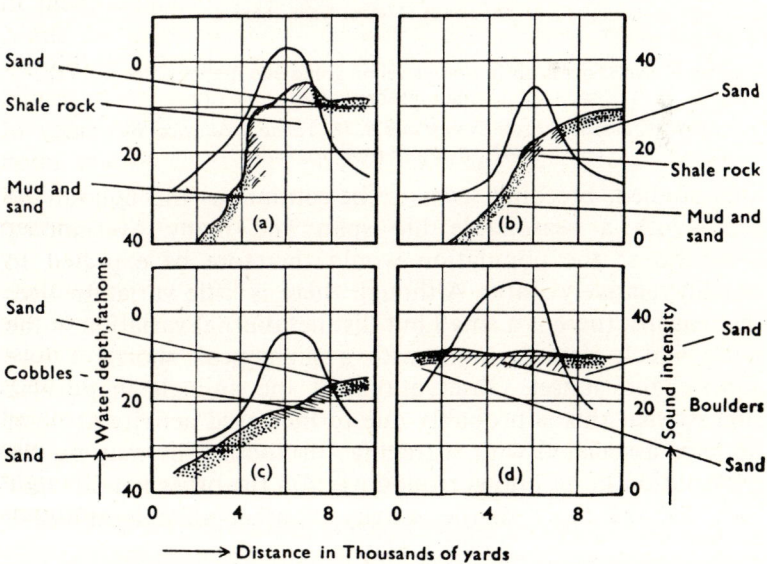

Sound levels and bottom profiles of traverses on above figure
a) Traverse AA' b) Traverse BB' c) Traverse CC' d) Traverse DD'

(Figure 55); the numbers are too small to give the continuous crackle, so that it may be assumed that the greatest numbers occur at depths above 30 fathoms. The noise is continuous over the shrimp beds, suggesting that the shrimp population is relatively constant and static. This is in agreement with the known behaviour of the animals. They tend to keep to their burrows, not migrating or moving great distances and, in the

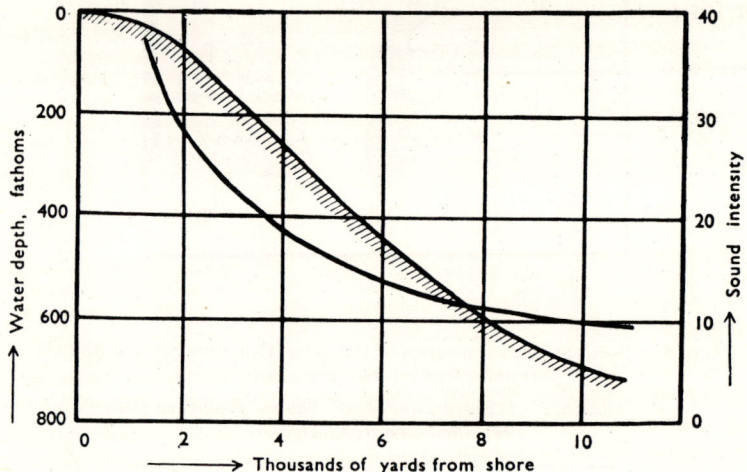

Figure 55
Diagram to show drop in noise level on going into deeper water from the high inshore values
(M. W. Johnson, F. A. Everest and R. W. Young, 'Biological Bulletin', 1947; Redrawn from original)

area studied, breeding seems to be continuous throughout the year—with a reasonable life span and many overlapping generations: the population would therefore be expected to remain relatively static. Although there is little variation over the seasons, there is a small but distinct diurnal variation in the noise which rises at night with a slight peak shortly before sunrise and sunset. Observations of the animals in aquaria suggest that this is probably due to increased activity at night perhaps associated with searching for food. This may be the explanation in natural surroundings; on the other hand it may then be due more to the activity of other animals at night,

THE USE OF SOUND WAVES

these in turn disturbing the shrimps, which are then stimulated to give the defensive snapping reaction.

These results indicate that on the whole the picture obtained —and obtained very quickly—by the listening technique has fitted the known behaviour of snapping shrimps; the method may, therefore, be applied with some confidence in situations where direct observations on the animal are lacking, sparse, or for some reason difficult to obtain.

3

SOME PROPERTIES OF THE WATER ITSELF

TEMPERATURE AND SALINITY

THE physical properties of fresh water are largely determined by two variables, temperature and pressure; in the case of sea water, a third variable must be added—the salt content or salinity. Temperature, salinity, and pressure, together determine the density of sea water, and a knowledge of its density is of fundamental importance in dynamic oceanography. Further, bodies of water are characterized by their salinity-temperature properties. In addition, information about temperature and salinity conditions and their variation enables deductions to be made about physical processes taking place at the sea-air interface, processes such as evaporation, precipitation and cooling; all of these are of great importance to physical oceanography. Temperature and salinity are no less important to marine biology. The temperature of the sea changes from the poles to the equator and the annual and seasonal variations differ widely from place to place. Just as on land there are arctic species and tropical species, species with wider and species with narrower temperature limits of distribution—so in the sea, and a knowledge of temperature conditions is required in studying problems of animal distribution. In the open oceans salinity changes are relatively small and their biological influence less marked, but in some regions, for example the Baltic Sea, salinity becomes greatly reduced and decreasing salt content plays an important part in restricting the spread of typically oceanic species into this area.

A knowledge of both the temperature and salinity of the water at a given point is required to calculate its density. A chemical titration of the water sample is usually made to determine the salinity and often, therefore, a sample of water is collected and the temperature taken at the same time—the temperature-recording apparatus being attached to the water sampler. Methods have recently been developed that do not require a sample for determining the temperature and salinity: these will be discussed later. The classical and still most accurate method consists essentially in letting down a suitable container into the sea—a container that can be closed at the required depth and to which is attached a thermometer, so that the temperature, either of the sample itself or of the surrounding water, is registered.

Surface temperatures

Surface temperatures are of considerable importance, since it is the sea surface which is in contact with the atmosphere. If the temperature at the surface only is required, the simplest method is to draw a bucket of water and take its temperature with an accurate thermometer, preferably one with a reservoir holding some residual water. The method is simple and the apparatus inexpensive, and many thousands of surface temperatures have been taken over the oceans, largely from merchant ships whose masters often co-operate in this work.

The method is subject to many sources of error which have been summarized by Dr Brooks of the U.S. Weather Bureau. The bucket is not likely to have the same initial temperature as the sea surface; the water sample is usually cooled by evaporation during hauling; the thermometer when inserted is seldom at the same temperature as the bucket; while the thermometer is resting in the bucket, further cooling or heating of the water may take place; when the temperature of the water is read, the thermometer may not have reached the temperature of the water in which it is immersed; if it is withdrawn for ease of reading, the temperature of the very small sample of water in the reservoir may change before the temperature has been observed; after the marking and numbers have become indistinct, errors of reading creep in; the thermometer itself may be inaccurate. Of course, many of these errors are negligible and many may readily be minimized; some would be quite insignificant if the measurements were made by scientists working on research vessels, but when these surface temperatures are taken by people whose primary job is not temperature investigation, errors such as those outlined must be considered in assessing the results. Dr Brooks, in a series of comparative experiments, found that about 34 per cent. of the readings differed from his standard reading by $-2°$ F. to $+7°$ F. The deviations, especially marked in cold winds, also appear to vary directly with wind strength and depression of wet bulb below sea temperature and inversely with the quantity of water drawn in the bucket; in fact, the chief source of error lies in any cooling of the sample during hauling.

To obviate many errors associated with the bucket method, Lt.-Comdr. Lumby of the Lowestoft Laboratory has devised a surface sampler which gives both the temperature and a sample for salinity estimation. It is simple, cheap, easy to use, gives reliable results, and can be handled from a merchant ship by untrained personnel. Essentially, a bottle and thermometer are towed in a torpedo-shaped container at the sea surface, the bottle being flushed out with the water. The apparatus consists of a head and body (Figure 56b) held together by a three-point bayonet fitting and safety catch. The bottle for the salinity sample is held firmly in a spring cup at the base and by a rubber washer bearing on the head at its upper end. The head carries a funnel-shaped opening which passes into a tube, the latter going well down into the sampling bottle. The thermometer, held at top and bottom in rubber glands, fits into a brass tube mounted within the main body. A celluloid insulating cylinder inside the body is separated from it at the sides by ebonite distance pieces and by an ebonite block at the base.

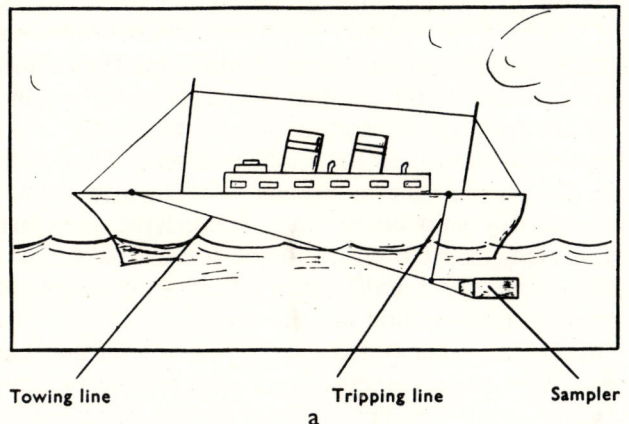

a) The method of towing

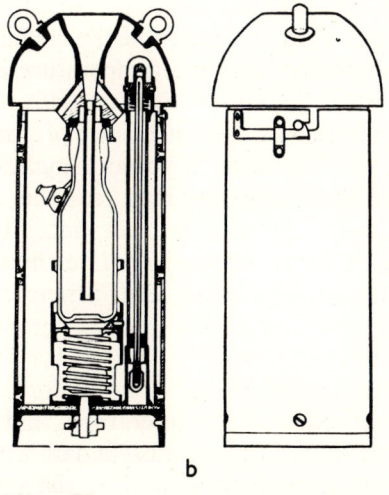

b) Section: the ordinary salinity bottle held by a spring underneath the conical mouthpiece and the thermometer in rubber glands at the side

Figure 56

The Lumby surface sampler
(J. R. Lumby, 'Journal du Conseil', 1928; Redrawn from original)

Towing bridles are shackled to eye-bolts on the head, and the usual method of towing is shown in Figure 56a. When the instrument is towed, water goes through the funnel piece in the nose and also enters the body by side tubes, whilst some passes into the sampling bottle and thence into the main body of the sampler; it escapes from this via ports under the head. The bottle is, therefore, continually flushed out during towing and the thermometer takes up the temperature of the water.

The bucket method and the Lumby sampler have now been replaced largely by thermographs, which give continuous readings with the ship under way. These have a temperature responsive element fitted either in the water intake pipe or in a pocket in the ship's hull and a recorder mounted in any convenient position on board.

Dr Church has summarized a large body of surface temperature data taken by these methods on merchant ships traversing the Atlantic Ocean. We may illustrate the results by considering average annual maximum and minimum temperatures along the New York-Cape São Roque route (Figure 57). The changes are quite striking; in the coastal waters there is a large temperature range, but further offshore we have the so-called slope water where the range is smaller and the temperature higher. The band of water constituting the Gulf Stream, with its high temperature, is clearly marked. Towards the centre of the ocean the temperature rises and the range becomes smaller until relatively constant conditions are attained in the North Equatorial Current. It should be emphasized that these are average values over long periods; we shall see later how great is the variability over short periods in the Gulf Stream region.

The insulated water-bottle

We must now consider the taking of temperatures and water samples at sub-surface levels. For this purpose a sampler must be designed which may be completely closed at the required depth so that the sample is not contaminated by water from any other level. Many types of samplers, usually called water-bottles, have been devised, but those that will function satisfactorily under working conditions at sea are few and of relatively simple and rugged design.

For moderate depths a bottle in which heat interchange is

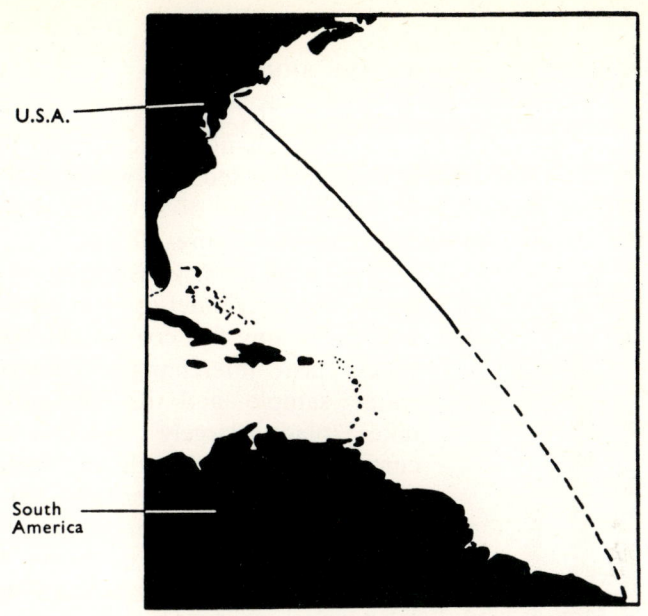

Route for surface temperatures given below

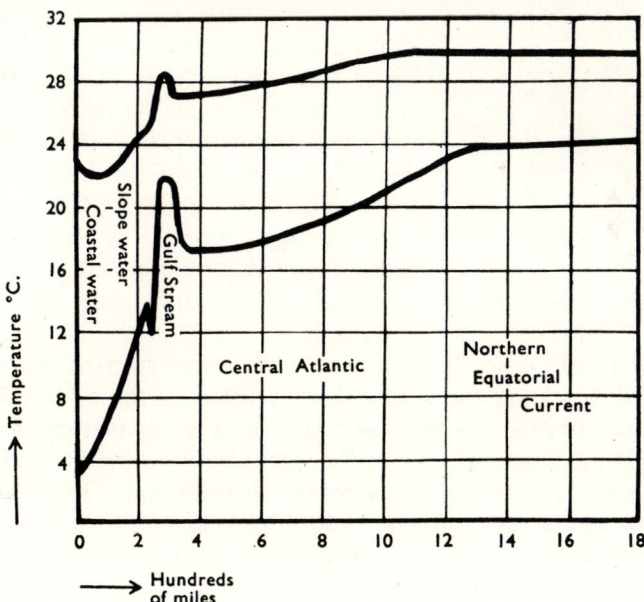

Figure 57
(*P. E. Church*, '*Association Internationale d'Océanographie Physique*', 1937; *Modified from original*)

reduced may be used and then the thermometer, mounted directly in the bottle, is read after hauling the sample to the surface. The bottle most frequently used and which will now be described is Knudsen's modification of a bottle originally designed by Nansen and Pettersson (Figures 58 and 59).

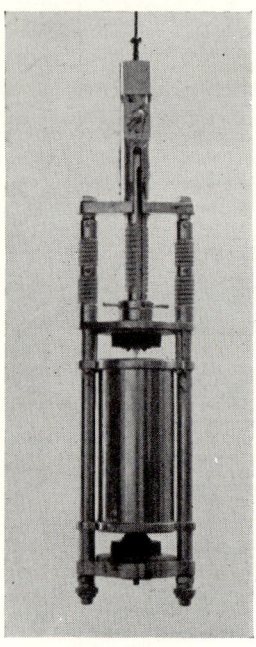

Figure 58

Insulated water-bottle in set position

The bottle itself consists of a thick-walled metal cylinder open at both ends and containing five coaxial thin-walled tubes; heat interchange between the water sample and its surroundings takes place largely by convection currents within the bottle and these are minimized by the coaxial cylinders which hinder any circulation. This cylinder, the body of the bottle, slides on two hollow pillars to which the fixed 'bottom' of the bottle is attached; the bottom plate, cushioned with rubber on its upper surface, has a valve for withdrawal of the water sample after the bottle has been closed. The lid of the bottle also slides on the hollow pillars and the thermometer, mounted through this lid, projects into the body of the bottle. The lateral guide pillars contain heavy springs, fastened at the bottom, and connected at the top with two wider pieces of tubing that slide over the inner pillars.

The bottle is set for lowering as follows: each of the rifled tubes sliding over the side pillars is lifted as far as it will go and then twisted, so that each slips into a bayonet catch that holds it up. The upper lid of the water-bottle is then raised (the pressure of the springs now being released) and a knob at the top, above the thermometer mounting, is caught by a hook and this holds it up. The two rifled tubes are then released from their bayonet fittings so that they press on the lid of the water-bottle and hold it down against the knob and catch with the full force of the springs. When the required depth has been reached, the bottle is allowed to remain there for 3 to 5 minutes, so that the thermometer may reach equilibrium and a 'messenger' (see p. 18) is then sent down the wire. The hook is released and the springs contract closing the bottle (Figure 59). The water in the tubes carrying the springs (with a combined pull on the lid of 16 kg.) damps the movement of the latter sufficiently to prevent any damage to the thermometer when the bottle is closed.

The run-off valve fitted to the bottom plate is kept closed by a spiral spring. When this spring is pressed upwards by the mouth of the bottle in which the water sample is being collected, the valve remains open and the water flows out; an air valve at the upper end of the water-bottle is opened to allow the water to run out freely. In strong currents heavily weighted extension legs may be fitted and these keep the bottle from drifting; such extension pieces also serve both to reduce the risk of damage if the bottle comes in contact with the bottom and to prevent contamination of the sample by mud.

A high degree of accuracy is necessary in the temperature measurements since small changes in temperature have relatively large effect upon the density. An accuracy of $\pm 0.01°$ C. is desirable and this can only be achieved with a well-made and accurately calibrated thermometer whose calibration is checked at regular intervals during use. It should be of small thermal capacity, so that equilibrium is quickly established. The scale must be open, easily read and have divisions every tenth of a degree. The marks should be etched on the glass and carried right round the thermometer so that errors due to parallax are more easily avoided.

The reversing water-bottle

When greater depths are being investigated, the time of hauling becomes quite considerable and the danger of heat exchange is increased. Further, the enormous changes of pressure in bringing the bottle up from, say, several thousand metres, results in adiabatic cooling. Under these conditions temperatures taken with an insulated bottle are not accurate. For deep-water work

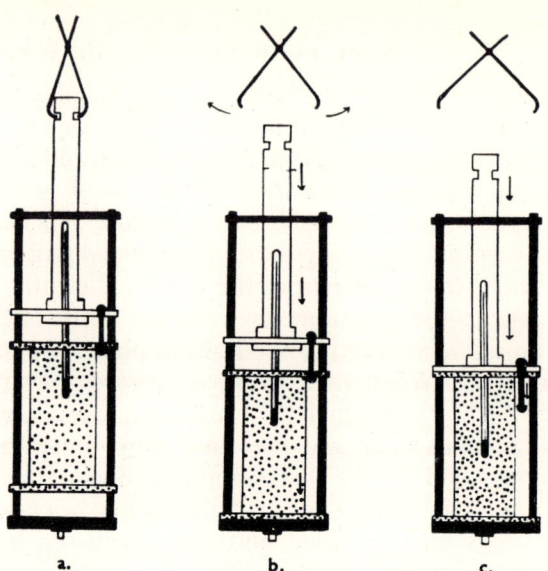

Figure 59
Insulated water-bottle. The closing mechanism (diagrammatic)
a) Raised in set position
b) Released bottle and lid free to fall
c) Closed; ready for hauling

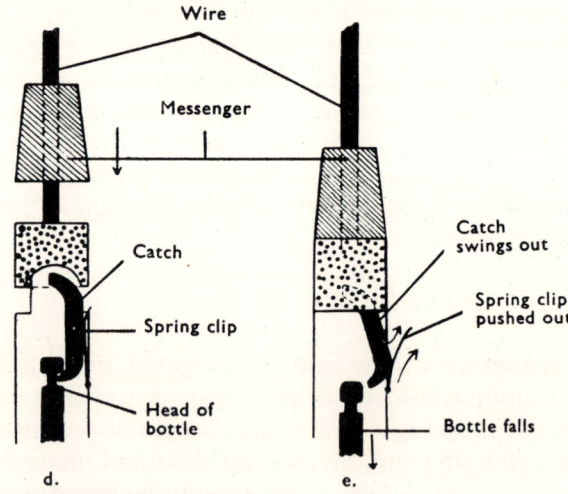

d) Set position with head held on catch, messenger about to strike and actuate lever
e) Catch released, bottle falling

SOME PROPERTIES OF THE WATER ITSELF

the temperature must be taken *in situ*. Thermometers of a special kind, called reversing thermometers, are attached outside the bottle, and the mechanism that closes the bottle also reverses the thermometers.

A reversing thermometer is essentially a double-ended thermometer with a large mercury reservoir at one end, and this is connected through a very fine capillary to a smaller bulb at the other end (Figure 60). Just above the large reservoir the capillary is constricted and branched with a small arm; above this, the thermometer tube is bent over into a loop from which it continues straight (over the 'registering' portion) and terminates in the small bulb. The thermometer is so constructed that when it is in the set position the mercury fills the reservoir, the capillary, and part of the bulb. The amount of mercury above the constriction depends upon the temperature, and when the thermometer is reversed by turning through 180° the mercury column breaks at the point of the constriction and runs down, filling the bulb and part of the graduated capillary, thus indicating the temperature at the time of reversal. The loop in the capillary, which is generally of enlarged diameter, is designed to trap any mercury forced past the constriction should the tem-

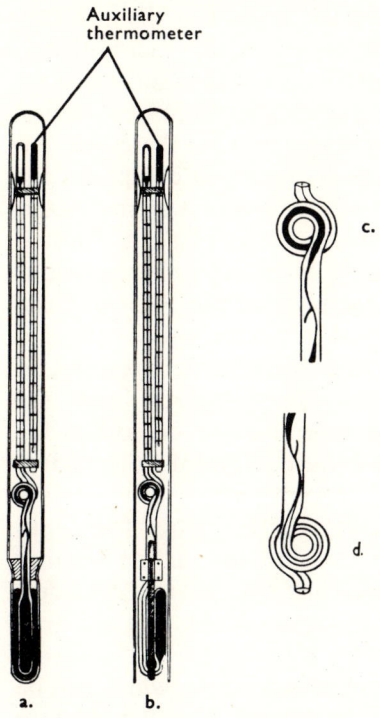

Figure 60

Reversing thermometers

a) Protected thermometer in set position
b) Unprotected thermometer in set position
c) Constricted part of capillary in set position
d) Constricted part after the reversal

perature be raised after the thermometer has been reversed. The thermometer readings are corrected for changes resulting from differences between the temperature at reversal and that of the surroundings when it is read, by reading at the same time a small standard type of thermometer, known as the auxiliary thermometer, mounted alongside the reversing thermometer. Because of the different compressibilities of glass and mercury, thermometers subject to great pressures give 'fictitious' values. The reversing and auxiliary thermometers are, therefore, enclosed in a heavy glass tube which is evacuated except for the portion surrounding the reservoir of the reversing thermometer. This portion is filled with mercury to serve as a thermal conductor between the surrounding water and the reservoir. However, this pressure 'error' is put to good use. A second unprotected thermometer, not enclosed in a strong glass tube and not, therefore, protected from the hydrostatic pressure, is fastened alongside the protected thermometers, and the difference in the readings can then be used to determine the depth at which the thermometers were reversed—the instruments being devised so that the apparent temperature increase due to hydrostatic pressure is about $0.01°$ C./metre. The depth obtained in this way is often of great value since in bad weather the wire to which the bottles are attached runs out at a considerable angle, and the amount paid out does not give the true depths of the bottles. If temperature alone is to be measured, the several thermometers are mounted in a frame, which can be reversed by means of a messenger. In general, however, a water sample for salinity is required at the same time as the temperature is taken and the thermometers are usually attached to a bottle, which itself may reverse within a frame or which may, together with the attached thermometers, turn completely over.

The Knudsen frameless bottle, in which the whole bottle turns over, has advantages over some other types, namely that it is lighter to handle and its closing device is very reliable; this sampler and its mode of action will be described.

The bottle is shown in the set position in Figures 61 and 62. It is clamped to the wire at the bottom, being free to swivel at the clamp, but kept in position by the wire itself, which passes,

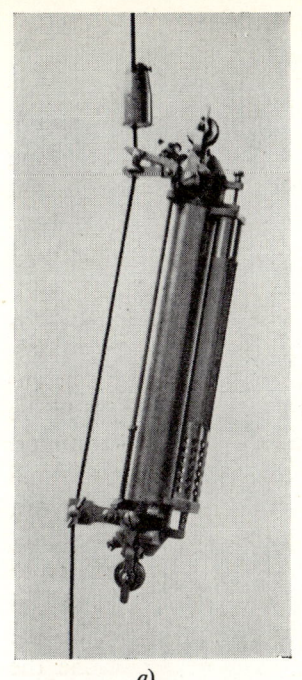

a)

b)

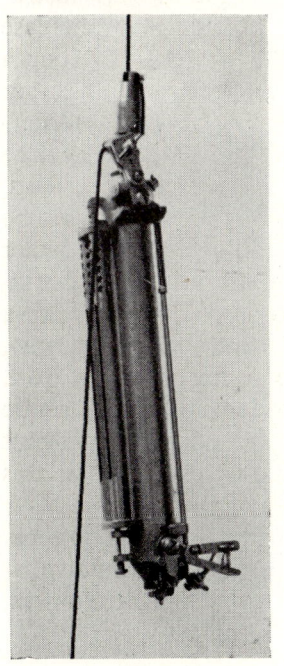

c)

Figure 61

Reversing water-bottle (Knudsen) in action

a) Messenger about to strike
b) Bottle closed and reversing
c) Completely reversed, messenger at end of fall

OCEANOGRAPHY AND MARINE BIOLOGY

at the top of the bottle, through a small stop attached to an arm and this, in the set position, also holds open the two movable lids. When hit by a messenger, the stop is forced away, the

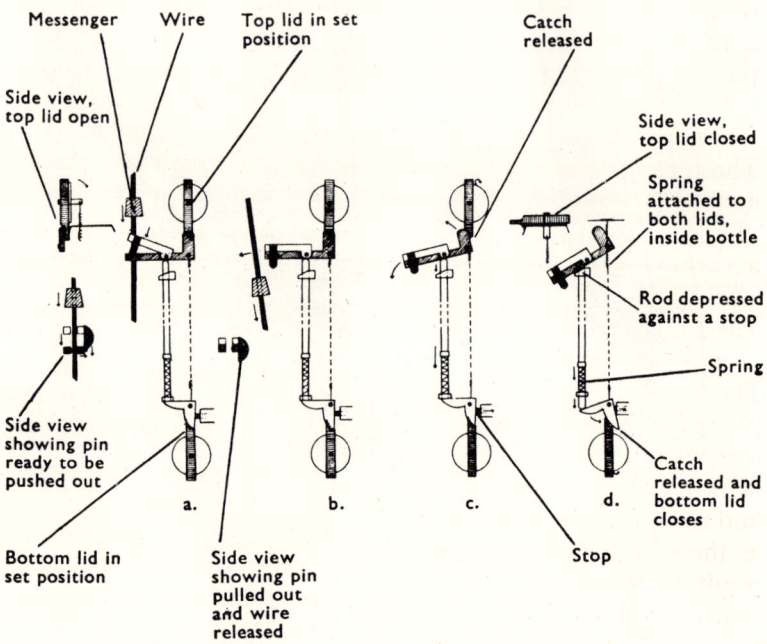

Figure 62

Knudsen reversing water-bottle (diagrammatic)

whole bottle and attached thermometers fall over and the latter are therefore reversed. At the same time the arm releases the catches (top and bottom are connected by a rod) holding open the bottle lids and they are then closed by the action of a stout connecting spring running through the centre of the bottle. The lids are well out of the way during descent so that water passes freely through the instrument. A second messenger can be released at the same time as the bottle is reversed and the former then travels down and closes, in exactly the same way, a lower bottle; a series of bottles may thus be used on one wire.

The thermometer and water-bottle are still the standard instruments for taking temperatures and for the collection of

SOME PROPERTIES OF THE WATER ITSELF

water samples for salinity estimations and chemical analyses. They give the most accurate results and are still widely used. However, since they are so time-consuming a vertical profile of temperature and salinity (or density) is only obtained at a relatively small number of points over a given area. As in plankton work, so in physical oceanography there is a demand for more closely spaced observations rapidly taken so that we may hope to make some approach to a synoptic picture. The instruments to be described next are all designed to give a more complete picture of the temperature and salinity structure of the oceans, but it must be stressed that, in general, some accuracy in the results has to be sacrificed.

The Mosby thermo-sound

The thermo-sound of Professor Mosby enables a continuous plot of temperature throughout a water column to be obtained and, unlike some other recording instruments to be discussed later, can be used to great depths. It is both simple in principle and robust in construction (Figure 63). A taut thin wire is used as the thermo-sensitive element and its change of length, as a result of thermal expansion or contraction, is recorded and utilized to measure the temperature.

This thermo-sensitive element, in the form of a length of fine brass wire, is attached to a frame at its upper end and to a stiff steel spring at its lower end. The frame, made of invar steel, with a negligible coefficient of thermal expansion, is strutted by cross-pieces and bolted to a head and base. The head carries a hole for a shackle and the base, besides taking the propeller mountings, also has a hole to which a weight may be shackled. The steel spring is attached at one end to the frame by a brass holder. The head-piece, where the brass wire is attached, is bent over, forming a support for a recording arm to which it is fastened. A 'pen' is soldered into the lip of the recording hand and a second hand attached to the frame carries a zero pen. The record is scratched on a small (6-cm.) smoked circular glass plate held on a table by two small springs, the table being turned (through a set of gears) by the propeller. The recording arms can be lifted out of contact with the smoked plate by a vane, but, as the instrument goes down, the pressure on this vane keeps the pens depressed and in contact with the smoked plate. The propeller is shielded by two brass guards and a simple, protective but not watertight, cover.

As the instrument is lowered, the propeller turns the table and the pen scratches a line on the smoked plate: the form of trace depends on the tension on the spring, which in turn is determined by the effect of temperature changes on the long brass wire. If temperature were constant the pen would trace a circle, but when it decreases, as is usual during descent, then the pen is drawn towards the centre. The traces are only very small (a movement of 0·15 mm. corresponds to 1° C.) and they are, therefore, measured under a microscope. The instrument is best calibrated by immersion in a water bath at a known series of temperatures. A rate of descent of about 2 metres per second is suitable. If the instrument is adjusted so that about 0·33° revolution of the plate corresponds to one metre depth, one full revolution of the plate will suffice for about 1,100 metres. Since, as already pointed out, temperature generally decreases with depth, there is little risk of confusion in the record with a second or even third revolution of the plate. As many points

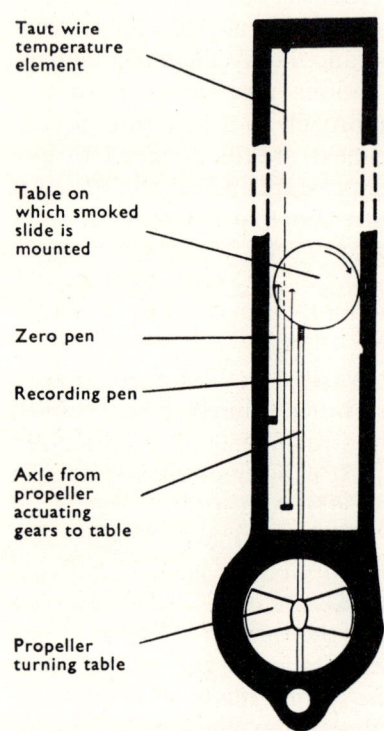

Figure 63
The Mosby thermo-sound
Diagram of mode of action

SOME PROPERTIES OF THE WATER ITSELF

as desired can be read off the record and the temperature-depth curve plotted.

Resistance thermometers and thermistors have been used for temperature measurement and conductivity cells for salinity. The instrumentation, however, is not simple, particularly if a recording device is incorporated into such equipment.

The Spilhaus bathythermograph and sea sampler

A new instrument, the bathythermograph (later combined with a water sampler), has been more widely used. This is a purely mechanical instrument, the principle being essentially that of the thermo-sound; a temperature-responsive element scratches a mark on a smoked plate which moves in response to depth. The depth response is, however, obtained by a pressure element rather than a propeller and the temperature-responsive unit is not a wire but a liquid in metal thermometer with a Bourdon spiral. (This replaces the bi-metal strip of earlier models, which were found to be very sensitive to vibration, and the records were, therefore, not very sharp.) The temperature element is exposed to the flow of water whilst the moving portion of the Bourdon spiral is entirely shielded. Effective heat transfer to the sensitive element is thereby ensured without transmitting vibration to the recording pen.

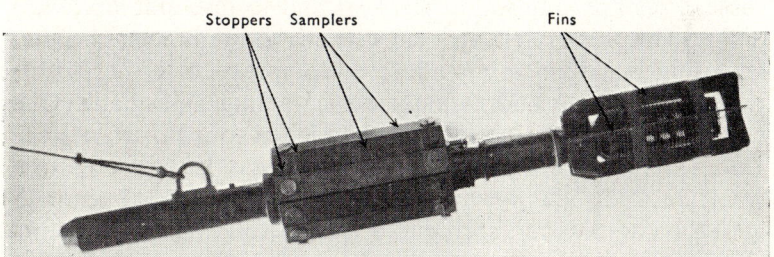

Figure 64
The bathythermograph and sea sampler

The main casing of the bathythermograph and sea sampler (see below) is a stout brass tube which houses the pressure element towards the rear end, the slide holder being fitted in the intervening space (Figures 64 and 65). A sliding brass sleeve fits over the main casing and three ports on the former coincide with three similar ports on the latter. By rotation of this sleeve, access to the slide holder and ejection of the slide is attained at two of the ports; the third port, when open, allows the pen stop to lift, and on closing causes the pen to fall back in position ready to mark the smoked slide. A heavy brass stream-

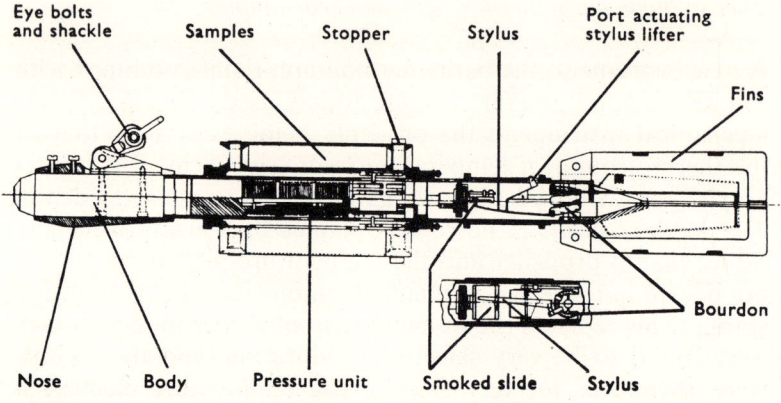

Figure 65
The bathythermograph and sea sampler

lined nose-piece and sleeve fit over the forward part of the casing; to the nose is fastened the bracket for the towing cable. Fins are attached at the rear end to give stability when diving. The pressure element consists of watertight metal bellows enclosing a carefully wound steel compression spring inside which is a piston and cylinder device acting as a guide. One end of the bellows is permanently attached to a brass endpiece within the body and the other to the slide holder that takes the small smoked slide on which the stylus marks; this is in contact with the compression spring and free to follow its movements (Figure 67). Thus as the pressure increases the slide holder is drawn under the writing stylus, which in the absence of temperature changes would then scratch a straight line (Figure 66).

The thermal element is a liquid thermometer consisting of a capillary and Bourdon tube, the latter being connected to the stylus (Figure 65); changes of temperature would cause the

SOME PROPERTIES OF THE WATER ITSELF

stylus to move in an arc across the smoked slide if pressure remained constant (Figure 66). The capillary tube is taken outside the main casing and wound on six slotted fins, where it is fully exposed to the surrounding water.

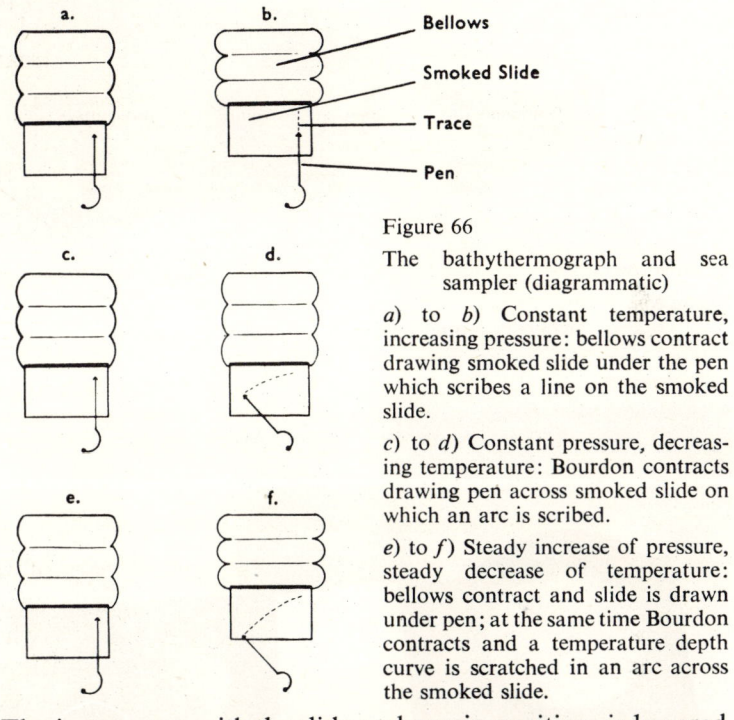

Figure 66
The bathythermograph and sea sampler (diagrammatic)

a) to *b*) Constant temperature, increasing pressure: bellows contract drawing smoked slide under the pen which scribes a line on the smoked slide.

c) to *d*) Constant pressure, decreasing temperature: Bourdon contracts drawing pen across smoked slide on which an arc is scribed.

e) to *f*) Steady increase of pressure, steady decrease of temperature: bellows contract and slide is drawn under pen; at the same time Bourdon contracts and a temperature depth curve is scratched in an arc across the smoked slide.

The instrument, with the slide and pen in position, is lowered over the side and towed at the surface for half a minute. The line is then paid out as fast as possible with the ship under way, and when approaching the required or maximum working depth the brake is slowly applied to the winch drum; after coming to rest, the instrument is hauled in. The slide will then have a temperature-depth record scratched on it. There may be one trace for the descent and one for the ascent; often the two are coincident. The slide is inserted in a viewer and read off against a calibrated grid (Figure 68); a photograph gives a permanent record.

Figure 67
Temperature unit and stylus of bathythermograph

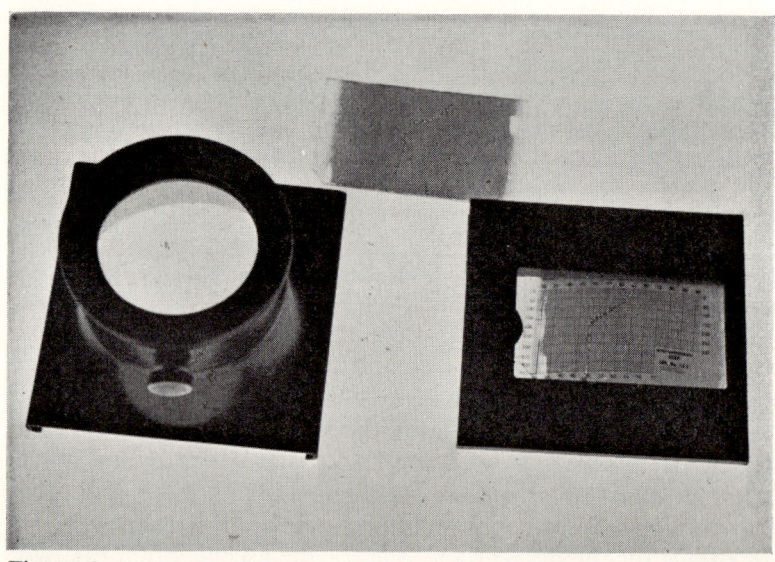

Figure 68
Holder and calibrated scale for bathythermograph slide. The slide is inserted into the holder behind the calibrated scale and read through the magnifying glass into the base of which the slide holder fits

SOME PROPERTIES OF THE WATER ITSELF

The bathythermograph gives a temperature-depth record. In its modified form samples of water are also taken at selected depths for salinity estimations. This is achieved by allowing the pressure element to actuate a series of triggers at given depths, these triggers closing the sample bottles. Both pressure

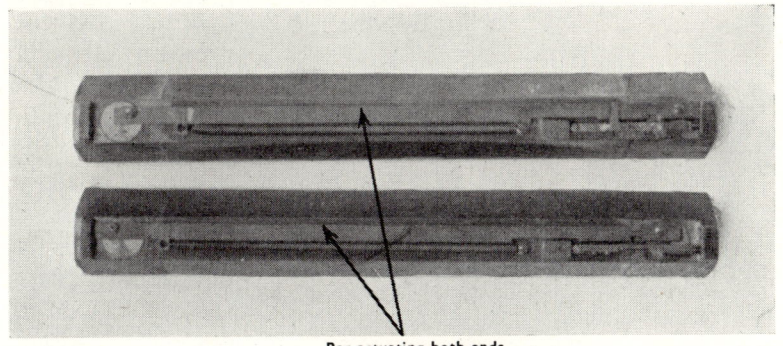

Figure 69

The sea sampler and bathythermograph
Two of the twelve sample containers

Figure 70
Enlarged view of Fig. 69. Note catch is at different distances from end to trigger at different depths. Near bottle is in set position

and thermal elements as well as the general construction are similar to the simpler instrument just described, but the main casing has twelve slots to enable the trigger mechanism of the bottles to be operated by the pressure element. The metal bottles are arranged radially around the case and each is provided with a closing valve at both ends. These valves are open in descent, during which a trigger associated with the valves of each bottle slides along a supporting strip of the sample operating unit, which is moved by the pressure element (Figures 69 and 70). During descent the bent end of the strip (trip) is pushed aside (Figure 71) as it passes the trigger, but on its return (during the ascent), when the trip comes against the bottle trigger the latter is now pushed to one side and the valves are closed by a spring mechanism, so enclosing a sample of water. The triggers are set

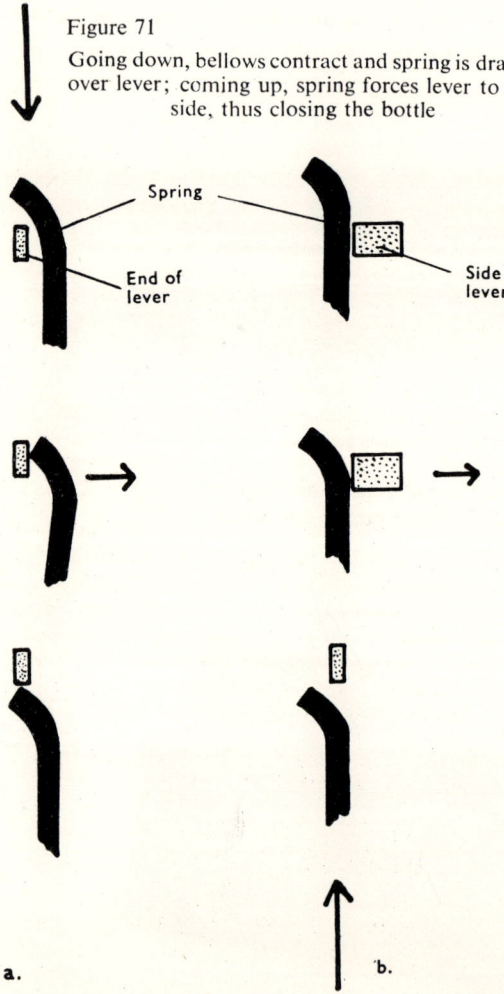

Figure 71
Going down, bellows contract and spring is drawn over lever; coming up, spring forces lever to the side, thus closing the bottle

at positions (Figure 72) in which they will be tripped in this way at selected depths.

This instrument has been intensively used in Gulf Stream studies and the results have clearly indicated features impossible

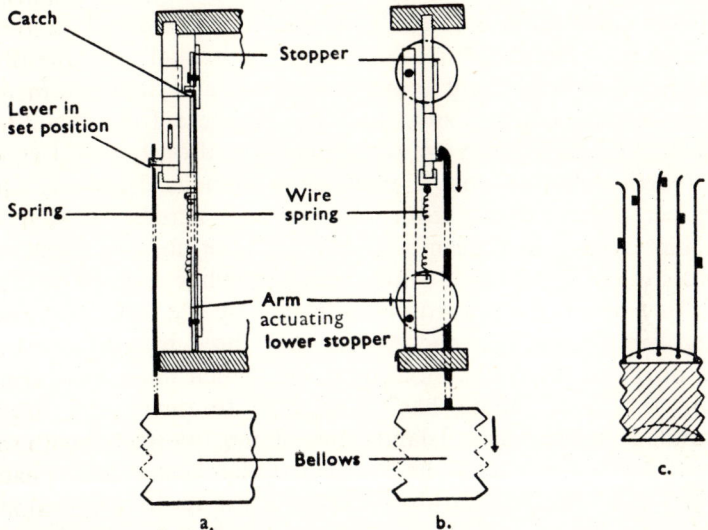

Figure 72
The bottle-closing mechanisms:
 a) Side view
 b) Top view
 c) Diagram showing levers set at various levels and springs on the bellows

to detect without such rapid methods of temperature and salinity estimation. Large eddies, entraining water from outside the Stream yet carried along with it, have been found; warm eddies, probably broken-off loops, have been detected many miles away.

CURRENTS

Direct ways of measuring currents may be divided into two groups—drift methods and flow methods and, although both are to some extent complementary, the form and type of information obtained varies with the instrument used. In planning work on currents it is necessary to realize the limitations of each method; the technique used must depend upon the kind of information required and upon the use to which it is to be put.

We shall first consider drift methods. The general drift of surface currents is known from the movements of floating objects and there are many interesting 'casual' observations which tell us something about their speed and direction in the open oceans. Glass floats, used by Japanese fishermen, and wrecked Chinese junks are sometimes found on the west coast of North America and it may be presumed, therefore, that the net current flows from west to east across the North Pacific Ocean. In December 1887 an enormous lumber raft consisting of 27,000 tree-trunks broke up while under tow from the Bay of Fundy towards New York. By the following June logs were found near the Azores and three months later to the north of Madeira. Floating wood, which must have come from the West Indies, has, from time to time, been found stranded on the Faroe Islands and mahogany trees have been found washed ashore on the south Greenland coast. In the West Indies is a climbing shrub remarkable for its long pods, and these are often found washed up on the shores of the Faroes, where they are known as 'goblin's kidneys'. These and many other examples of tropical plants cast up on the west coasts of Europe all point to a general drift of surface water across the North Atlantic in a north-easterly direction. Again, Siberian timber frequently drifts across the Polar Basin to the north of Spitzbergen or to the east Greenland coast, and it was on this sort of evidence that Nansen planned his famous *Fram* Expedition; he deliberately froze his ship in the pack off the New Siberian Islands and allowed it to drift with the ice; it averaged about a mile a day as far as Franz Josef Land. The movement of icebergs gives very reliable indications of currents, since, having a large proportion submerged, their wanderings are little affected by wind. Bergs, originating in the Arctic, are regularly found in waters to the south-east of Newfoundland and indicate the existence of a south-going current in this region.

Bottles containing postcards have often been put out from

SOME PROPERTIES OF THE WATER ITSELF

exploration vessels and recovered after what must have been long journeys. A glance at a map shows that in the southern oceans—south of the great continents—there is an unobstructed waterway round the earth, and there is reason to believe that a bottle put out from a German barque about half way between Kerguelen and Tasmania, and which was recovered, 2,447 days afterwards at Bunbury in Western Australia, had probably travelled some 16,000 miles round the world at an average speed of some 7 miles a day.

Drift bottles and drift methods

In view of the above, it is not surprising that systematic experiments are regularly carried out by means of so-called 'drift bottles'. In all drift-bottle experiments there are, however, difficulties both of technique and interpretation. For example, it is essential to ensure that, as far as possible, the drift bottles do not 'sail' through the water, being carried forward by the direct action of the wind. In any analysis of the results it must be remembered that the bottle may not have followed a straight course from its point of liberation and that it is not known how long the bottle may have been lying in the position where it was found. Further, the returns of these cast-up bottles will be clearly dependent upon the frequency with which the coast-lines are visited; some may find their way to isolated shores. However, drift-bottle methods are cheap, easy, and rapid; bottles can be put overboard from a vessel in weather that would make any other type of current-work quite impossible. From the early days of research in oceanography they have been used systematically on a large scale by many people and have yielded valuable information.

The bottle (Figure 73d) is so ballasted that it floats with its neck just awash and still further to reduce the direct effect of wind, is made long and narrow. They float in the surface layer until cast ashore on a beach or, occasionally, are taken by fishermen in their nets. Each bottle contains a 'break this bottle' slip and a postcard (Figure 73a, b, c) printed in several languages and the finder is asked to fill in and return the latter: a small reward is usually offered. The results from such experiments only give a picture of the surface drift—the upper layer of water

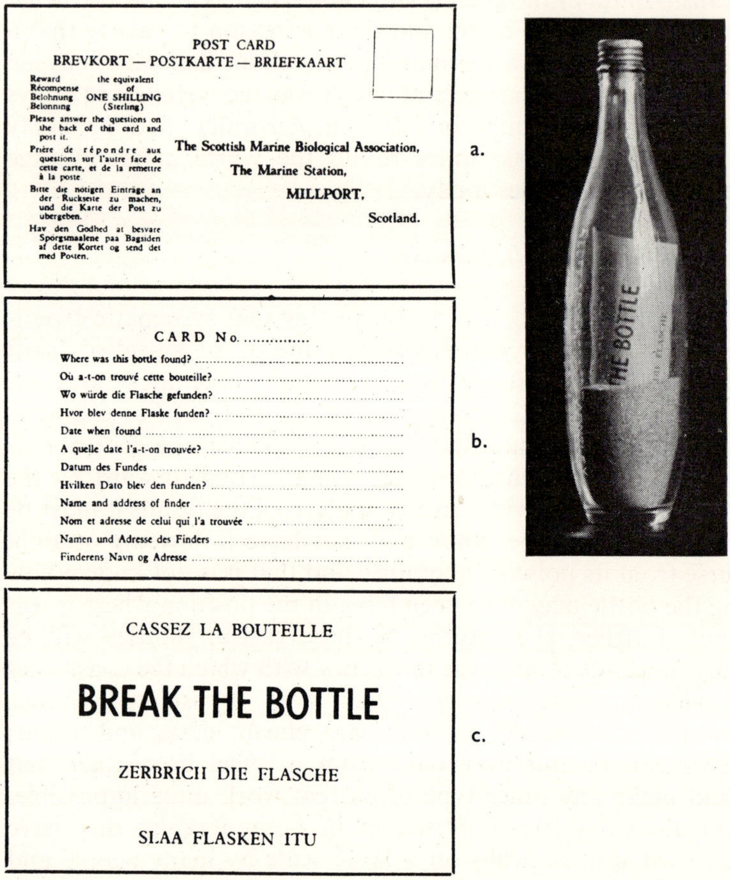

a.

POST CARD
BREVKORT — POSTKARTE — BRIEFKAART

Reward / Récompense / Belohnung / Belonning — the equivalent of ONE SHILLING (Sterling)

Please answer the questions on the back of this card and post it.

Prière de répondre aux questions sur l'autre face de cette carte, et de la remettre à la poste.

Bitte die notigen Einträge an der Ruckseite zu machen, und die Karte der Post zu ubergeben.

Hav den Godhed at besvare Sporgsmaalene paa Bagsiden af dette Kortet og send det med Posten.

The Scottish Marine Biological Association,
The Marine Station,
MILLPORT,
Scotland.

b.

CARD No. ....

Where was this bottle found? ....
Où a-t-on trouvé cette bouteille? ....
Wo würde die Flasche gefunden? ....
Hvor blev denne Flaske funden? ....
Date when found ....
A quelle date l'a-t-on trouvée? ....
Datum des Fundes ....
Hvilken Dato blev den funden? ....
Name and address of finder ....
Nom et adresse de celui qui l'a trouvée ....
Namen und Adresse des Finders ....
Finderens Navn og Adresse ....

c.

CASSEZ LA BOUTEILLE

BREAK THE BOTTLE

ZERBRICH DIE FLASCHE

SLAA FLASKEN ITU

d.

Figure 73

Drift bottle

a) *b*) Postcard to be returned by finder
c) Break the bottle slip
d) Bottle with card and slip, loaded with sand

which is very much under the influence of wind. Attempts have been made, therefore, to use deeper-riding bottles: these are buoyant bottles similar to those just described but have

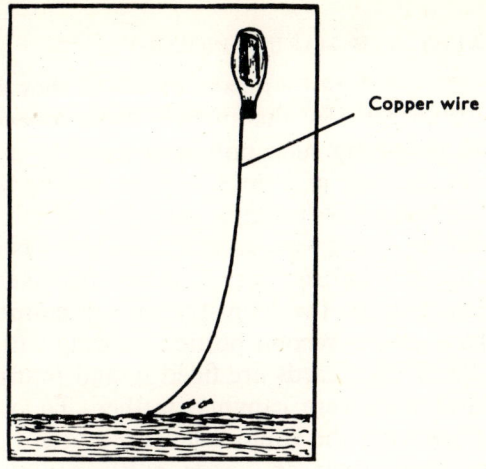

a Bottom trailing bottle

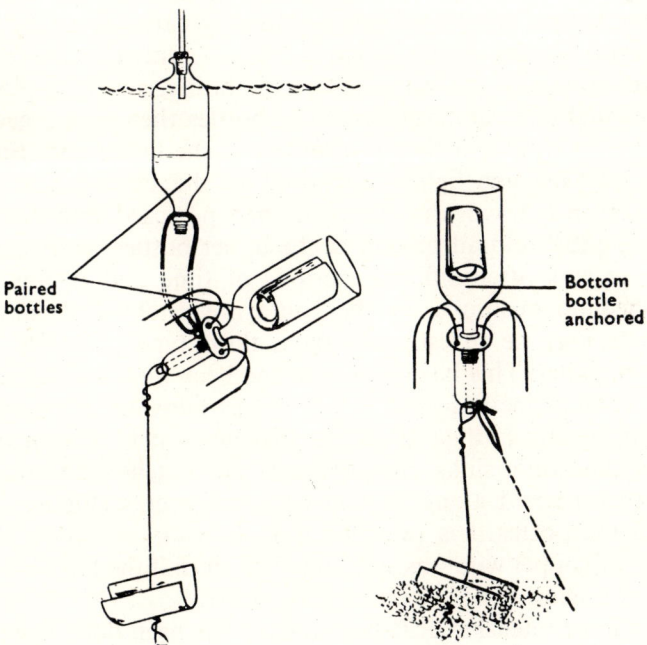

Figure 74
b Double drift bottles
(*Redrawn from figure in* 'Fisheries Notice, No. 16'; *British Crown Copyright*)

OCEANOGRAPHY AND MARINE BIOLOGY

attached to them a drag by means of which they are made to ride some distance below the surface. Since one of the chief difficulties in preparing such bottles is to make a satisfactory attachment of the drag and since, if it eventually breaks away, the bottle then behaves as a simple surface bottle, it is desirable to know whether the drag was still attached when the bottle was picked up; this information is, therefore, asked for on the postcard. Scientists of the Ministry of Agriculture, Fisheries and Food have used a second bottle as a drag with a postcard in each bottle; if both cards are filled in and returned, there is no doubt that they were caught together. To overcome the difficulty of knowing the route followed by the bottle, paired bottles have been used; an upper bottle contains acid that eats through a stopper and drops the lower bottle *en route*. The upper bottle has a capillary tube at the top and a metal stopper below, the latter being gradually eaten away by acid contained in the bottle (Figure 74b). Eventually the metal stopper collapses and water enters the bottle, air being pushed out through the capillary leak. The bottles then sink together, and since a simple anchor is attached to the lower one (Figure 74b), they become lightly anchored to the bottom and may be taken in a trawl and the contained postcard returned. By adjusting the amount of acid in the upper bottles, pairs can be made to sink after different periods of time and in this way intermediate positions on the route are established. Bottom currents may be investigated by a third type of bottle—the bottom trailer. This has weights in the neck and is so adjusted that, when properly sealed, it is only just buoyant; a long piece of copper wire is then added so that when it is thrown overboard the bottle sinks until the wire tail touches the bottom; it is then carried along by any bottom currents (Figure 74a). When the postcard is returned it is necessary to know what length of copper wire was still attached, since if the tail were lost it would again behave as a surface bottle.

The most intensive drift-bottle work has been done from the Ministry of Agriculture, Fisheries and Food Laboratory at Lowestoft and the Scottish Home Department Laboratory at Aberdeen in their work on the north-eastern Atlantic (particularly the Faroe-Shetland Channel), the North Sea and the English Channel. In all these regions there is a good chance of bottles being recovered from the bounding shores, and the intensity of fishing gives a reasonable recovery of bottom bottles. In 1920, for example, 68 per cent. of 9,550 surface bottles and 35 per cent. of 9,525 bottom-trailing bottles, put out by Lowestoft workers, were recovered. In 1929, bottles liberated at the western end of the English Channel (except those stranded locally) were found to travel rapidly up Channel, across the North Sea, and many arrived in the Skagerrak after a journey of some 700 miles at a minimum rate of 6 miles per day (Figure 75). It is significant that over this period there was for some five and a half months an almost uninterrupted predominance of south-westerly winds in this area.

Recently, plastic envelopes containing cards have been used instead of drift bottles; these have the advantage of being small and easily carried to the point of liberation. Unless specifically weighted or made with a weighted plastic tail they do, however, give only the drift of the very surface skin.

Figure 75
Routes taken by drift bottles
(*Redrawn from figure in 'Fisheries Notice, No. 16'; British Crown Copyright*)

Near land, drift methods may be easily elaborated. The movement of a buoy, or a coloured patch of water, may be followed for long periods and over quite considerable distances by following its course by ship and positioning it with respect to land bearings. This method may even be employed in the open ocean, using additional fixed buoys, and such a technique has been greatly facilitated by the development of accurate navigational aids such as Decca and Loran. Radio buoys giving out continuous signals have been used for the same purpose; they are, however, expensive.

Flow methods: the Ekman current meter

In flow methods, used from an anchored ship, the instrument is kept stationary and the current measured by allowing it to turn a propeller or to exert pressure on a movable vane. With this type of technique observations can be made at a series of depths and the speed and direction determined. The main disadvantages are that the apparatus requires constant attention and the records refer only to the limited period during which the observations are made. When anchored in deep water, difficulties also arise from the ship's motion. In most of these instruments a vane orientates the machine in the direction of the current and a mechanism is provided for recording this direction, relative either to the magnetic meridian or to the fixed direction of a bifilar suspension. Instruments that measure the pressure do not need a vane, since the deflection of a pendulum may be used to measure direction. The advantage of a propeller or cup instrument is that, in general, a linear relation exists between the number of revolutions per minute and the current speed, and this type of instrument has been most commonly used. It is not reliable, however, for current speeds of less than 2 cm./sec. The use of a compass to determine direction has the disadvantage that near a steel vessel the needle is influenced by the ship's magnetism; compass-containing instruments should not be used at depths less than 50 metres. Propeller and cup instruments both suffer from the further disadvantage that, unless adequate precautions are taken, they are easily fouled by passing objects such as seaweed.

There are many different forms of current meters employing these general principles. The various instruments reflect in part the period of invention, for example, the availability of ancillary electrical or electronic equipment, in part the bent of the inventor—mechanical or electrical—and in part the parti-

SOME PROPERTIES OF THE WATER ITSELF

cular purpose for which the instrument was intended. The simplest method for an accurate direct current reading is undoubtedly to allow the current to turn a propeller and in some way to record the number of turns in a given period of time, the direction being indicated by reference to a compass. We shall not describe all the various modifications of this technique, but rather illustrate the basic principles involved by reference to the well-known instrument of Ekman (Figures 76, 77).

The main body of the instrument (Figure 76a), mounted on ball bearings, is free to swing round its vertical axis in response to a current. If the direction of weak currents is to be recorded accurately, the plane of symmetry of the meter must be vertical, and the instrument is, therefore, carefully balanced. The tail orientating it into the current is a simple metal plate attached to the body by two brass tubes. An 8-bladed propeller is mounted inside a strong protecting ring to which two front shutters are hinged. Slow-running propellers, that is, with a high pitch, have been found to give the most accurate results. A set of levers and catches are arranged so that when the instrument has been lowered with the front doors closed and with the propeller locked, a first messenger will both open the doors and release the propeller and, after allowing to record for a known time, a second messenger will then lock the propeller. This is achieved as follows (Figure 76a). Inside the protecting ring there is a 'stopper' which moves vertically and which is connected with a lever so that the propeller is released or arrested as the right arm of the lever takes a higher or lower position. The lever is raised by a spiral spring and is kept in the set position by a catch in the shape of a twin spring on an arm resting against the closed upper shutter, which is in turn locked by the lever. The instrument is lowered into the water closed, with the double spring pressed to the right (towards the wire) and held by a ratchet. When hit by the first messenger the lever releases the shutters; these fly open by the force of the strong spiral springs and the lever immediately takes its upper position. At the same time the ratchet releases the twin spring, which now resumes its natural position to the left, catches the lever and locks the propeller when it is pressed down by the second

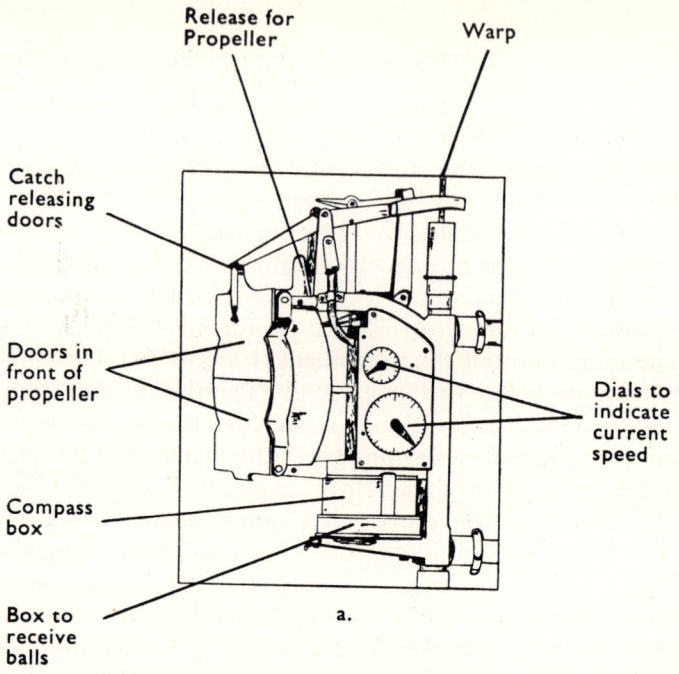
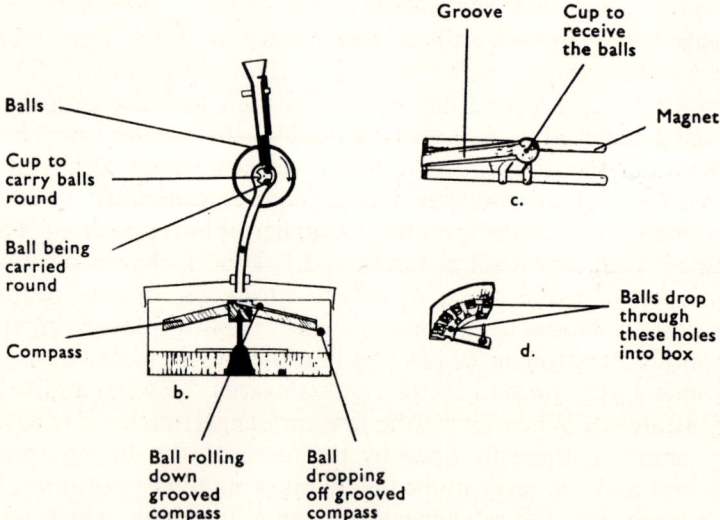

Figure 76   The Ekman current meter

a) Diagram of front part (without fin).   *b*) Diagram of balls feeding on to compass.   *c*) Compass with ball receiver and groove.   *d*) Compass mounting and holes through which balls pass to box

Figure 77
Current measuring—Ekman meter
(*Courtesy of the Ministry of Agriculture, Fisheries and Food, Lowestoft*)

messenger. Since connection between lever and stopper is indirect and elastic, the impact of the messenger is not transmitted to the blades. The propeller shaft leads, via gears running in stainless-steel bearings, to a set of recording dials, and these indicate the number of propeller revolutions. It is necessary to know the relation between propeller revolutions and water velocity. This is usually obtained from tank tests; velocity is calculated from the time and length of a tow and, knowing the number of revolutions of the propeller, the instrument may be calibrated.

Direction of the current is recorded mechanically in a very ingenious way. A compass with a grooved needle (Figure 76c) is mounted on the instrument and balls are dropped on to it and carried along this groove into a box fixed to the frame. Since this is orientated into the current, its direction with respect to magnetic north (as indicated by the compass) may be determined. A vertical brass tube serves as a magazine for the balls and its 'bottom' is formed by a three-holed cylinder carried on the axle of the worm wheel driving one set of the current dials (Figure 76b). As this cylinder passes the magazine a ball drops into one of the holes when the latter becomes aligned with the magazine base. The ball is carried round until it comes opposite a lower guide tube, when it passes down into the actual compass box. Here a system of magnets runs on agate bearings and the frame of the magnets carries a trough into which the ball rolls. The lower part of the compass box, rigidly attached to the swinging frame, is divided by means of radial partitions into thirty-six compartments each with a hole opening into an equivalent compartment of the lower storage box (Figure 76d). These compartments and not the compass box itself finally receive the balls.

This instrument gives only one result at each lowering and although in shallow water, or even in water of moderate depths, this is inconvenient, it is not too time-consuming. For deep-water work it would be extremely laborious to take a large number of readings, and Ekman therefore devised a more complicated instrument on the same basic principles. In addition to those dropping into the compass box, balls were

SOME PROPERTIES OF THE WATER ITSELF

guided into slots by the propeller dials. Up to forty-seven messengers could be sent down and, by using numbered balls, forty-seven consecutive readings could be obtained without bringing the instrument to the surface. After activating the instrument the messengers divide into two parts and fall into a receptacle hung below the instrument. The great advantage of these instruments, apart from being robust, is that they are entirely mechanical; no extra wires are required and there are no complicated electrical or other devices to go wrong.

Continuous-recording current meters

Electrical and photographic methods of recording both velocity and direction of currents have been used to obtain more continuous observations and propeller instruments employing such devices have had varying degrees of success and popularity. In one of the earliest electrical machines, devised by Rolf Witting—a Finnish hydrographer—the propeller turns an eccentric disc that actuates a forked lever working against a magnet. When the lever is raised, it lowers the magnet frame which serves as a key closing an electrical circuit by providing contacts between an inner solid and an outer segmented contact ring. One conductor is connected to the inner ring, and the other to the segments by resistances of different magnitudes. The resistance in circuit depends, therefore, upon which segment is touched by the outer ring of the magnet frame and, by having a milli-ammeter in circuit, orientation with respect to magnetic north may be determined. Since the propeller actuates these contacts, once the instrument has been calibrated, the current speed may be calculated by the number of contacts per unit time.

Photography is used to record in a current meter designed by Pettersson. A propeller turns a disc with a transparent section and the magnet carries a smaller similar and concentric disc. Every half-hour the positions of the discs are photographed. A clockwork mechanism advances the film and actuates a battery supply to an electric light. This instrument was designed to be suspended from a buoy and to work continuously for two weeks.

The most recent instruments rely on electromagnetic induction as the link between propeller and recording equipment, and since the signal generated in this way may be recorded continuously, continuous current measurements are possible. The link involves no moving parts. To form this link the propeller is fitted with a small strong permanent magnet at the tip of one or more blades and a small coil of wire is housed on the propeller guard. At each turn of the propeller, as the magnet passes the fixed coil a momentary pulse of induced electromotive force is generated, and current flows in the closed circuit (Figure 78a). The signal is transmitted via a two-core cable on which the instrument may be lowered (Figure 78b). The pulses may be counted, integrated or recorded. It should be noted that the energy of the electrical signal is derived from that given to the propeller by the passing water. Intermittent magnetic drag together with propeller-bearing friction usually places minimum detectable speeds at about 0·15 knot.

A count of the number of rotations may be obtained in several ways. The propeller rotation signal may be registered directly on a recording milli-ammeter (Figure 79). It may also be detected visually by a slow-response micro-ammeter, in which case very low current velocities are indicated by separate kicks on the instrument, while high velocities are recorded as a steady current value. The signals may also be counted by acoustic methods using a pair of headphones.

In the model designed by Dr von Arx for deep-water work, the instrument (Figure 78c) has a barrel designed to withstand the pressure at 300 metres. Within the pressure case is a magnesyn compass transmitter to report direction to the deck, a Bourdon-operated micro-torque potentiometer to give instrument depth, and a pick-up coil with selenium rectifiers to supply a micro-ammeter which indicates current speed. In order to smooth the micro-ammeter action, the signal impulse rate of this particular model is increased by mounting several smaller magnets separately on the tip of each propeller blade rather than a single large magnet on one blade. This improves the

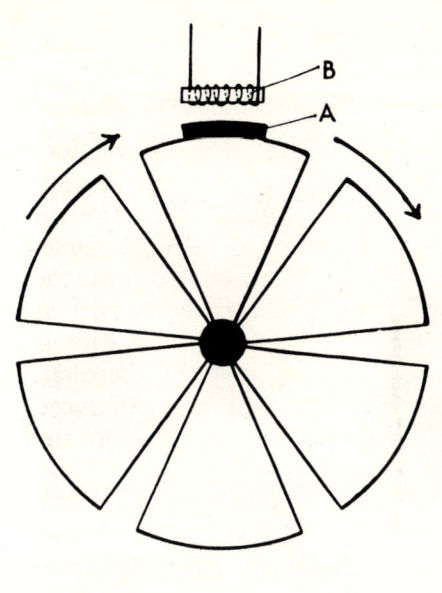

a)

b)

c)

Figure 78

The von Arx current meters

These rely on electromagnetic induction

a) As the magnet A on the propeller blade passes the coil B a surge of current is induced in the latter and can be observed or recorded on deck

b) Simple model. No direction indicator

c) Direction indicator (magnesyn) in pressure case: meter orientated into current by Garbell-type fins for short instruments

(*Courtesy of Dr W. S. von Arx*)

Figure 79

Measuring currents in the Mediterranean with the von Arx Model II current meter. The instrument is suspended from the small boom in the background. The deck unit showing current speed, direction and instrument depth is being read by Dean F. Bumpus, Eugene Krantz and Second Officer Donald Fay

(*Courtesy of Mr D. M. Owen*)

micro-ammeter action and its response at low current speeds, but it removes the possibility of counting individual propeller revolutions at the lowest current speeds. Dr von Arx has built models of this instrument both with conventional fins and also with the new type of Garbell fins (Figure 78c). The latter, developed in connection with modern aircraft, have a high directional sensitivity even on a short instrument with its greatly reduced rotational inertia.

SOME PROPERTIES OF THE WATER ITSELF

It has been pointed out that, particularly when measuring currents in deep water, the motion of a ship even at anchor must be taken into account. At a given instant one really measures the run of the current relative to any movement of the ship or, more precisely, to any movement of the current meter—for the latter does not follow exactly that of the ship. Even when firmly anchored, two components of motion are important. Under the combined influence of current and wind a ship will move round a central position, thus executing a pendulum movement within a sector whose point is formed by the anchor: it will in fact yaw, and this is the principal motion. Further, as the resistance offered by the vessel to wind and current varies with time, the ship rides to and from the anchor itself. The disturbing effects are, of course, greatest when weak currents are being investigated and when the records are being made from very great depths. In areas where the currents are strong and the depth small, the effect can often be neglected. Various attempts have been made to overcome these difficulties, for example, by using more than one anchor. This is time consuming and not always very successful. When conditions are unfavourable and particularly when they are changing suddenly, current measurements obtained in this way may be unreliable. A current meter fixed to a tripod which rests on the bottom has also been used for measuring bottom currents.

One of the great advantages of continuous recording instruments such as those developed by von Arx is that observations can be made during those phases of an anchored ship's motion when acceleration is great but velocity small. Preliminary observations allow incoming signals to be separated into ship's motion and water motion. Once the rhythm of the ship at anchor has been found, the technique allows observations to be made at predictable intervals. At the end of each yaw there is a period when the ship swings off on the opposite tack and during swing a certain point in the ship's length is temporarily at rest with respect to the bottom. The location of this point is usually well forward of midships and varies with the size and shape of the ship. If the current meter suspension is led over the rail into the water at this centre of rotation, the upper end of the current-meter cable will be held at rest with respect to the bottom for a matter of a few seconds or minutes, depending upon the particular ship. During this time measurements can be made with some degree of reliability provided it is long enough for the current meter to have come to equilibrium. At equilibrium during these stationary intervals the indicated current speed is minimal; both the indicated rate of change of direction and the indicated depth are maximal. The speed observed during the motionless period and the average observed on the preceding and succeeding yaws is then taken to be the true current speed at that time. Since this technique permits the ship to be at anchor, it is possible to make reasonably reliable measurements of current velocity at all depths where a single anchor will reach and hold.

Very recently two further methods have been used in attempts to eliminate the difficulties associated with anchored ships when measuring deep-water currents. In one a buoy, balanced to stay at the required depth (the density of the water at that depth being known), contains a series of explosive charges which are set off at intervals. The train of sound waves from these charges is transmitted through the water and may be picked up by directional hydrophones at shore stations. The position of the buoy can, therefore, be determined at every explosion and its course plotted. An alternative method has been to enclose in a buoy a simple sound source rather like a modified transmitter of an echo-sounder. Sound pulses are continuously sent out, and these can be picked up by hydrophones on a moored ship and the positions of the buoy plotted.

Long-period current measuring instruments

All the methods described so far are capable of giving current speed and direction at any instant—but they are essentially methods requiring a research ship. In shallow coastal waters, where large tidal streams are super-imposed on permanent or semi-permanent currents, the results, while of interest, are often difficult to use in some forms of biological work, for example, that concerned with the dispersal of larvae and the movement of planktonic animals. For this type of investigation an instrument is required which will give the residual or net current over long periods—and which, therefore, requires little servicing: this aspect of current measurement has been given particular attention by Dr J. N. Carruthers of the National Institute of Oceanography. He has stressed that a continuous programme of current observations cannot be carried out by research vessels alone and his instruments have, therefore, been designed to be worked from lightships. Observations must, of course, be made at a number of different places and if these are carefully chosen the results give an indication of water movement over a wider area.

A simple current-measuring device, the Jacobsen current meter, for use on lightships or similar vessels, had indeed been in use for some time prior to Dr Carruthers' work, but readings had to be taken regularly by the crew. With this instrument current speed is measured by the tilt produced on a plate and the direction determined by reference to the ship's head. No magnet is involved and steel ships may, therefore, be used. There is, however, no continuous record of either current speed or direction; the instrument must be read at intervals.

Fastened firmly to the ship's rail is a strong bracket secured at the inboard end by two stout clamps, the jaws of which are adjusted by bolts under the rail (Figure 80). The bracket carries a flat iron plate swung in gimbals, and on the upper surface of this plate are two circular spirit levels. In each of these the moving element is a bubble of air. From the underside of the plate, upon which the two levels are mounted, an iron tube projects downwards from the centre for a short distance. This tube is screwed into the plate at its upper end and above it is a small hole bored through the centre of the plate. The hole is exactly midway between the levels. A davit lined up with these carries a small hand-winch that can readily be locked when desired. From this winch a steel wire leads over a measuring block, passes down through the central hole in the

OCEANOGRAPHY AND MARINE BIOLOGY

Figure 80
Jacobsen current meters in use
(*Courtesy of the Ministry of Agriculture, Fisheries and Food, Lowestoft*)

SOME PROPERTIES OF THE WATER ITSELF

plate and on through the iron tube. The free end is fixed to an open cylinder. The current, acting on the cylinder, takes it away at an angle and the levelling plate is tilted (Figure 81).

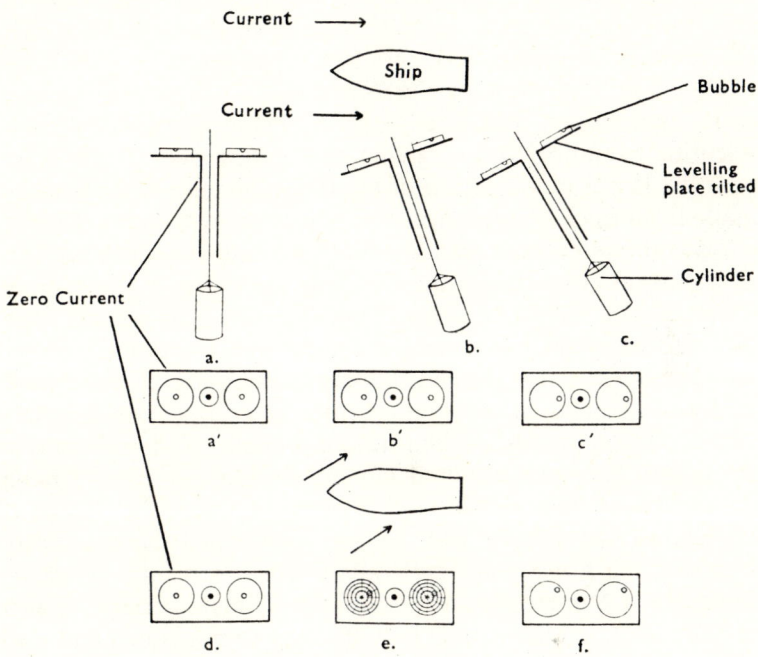

Figure 81

The Jacobsen current meter (diagrammatic)

a) to *c*) Ship heading directly into current
    a) No current
    b) to *c*) Increasing current } Side views
    a') to *c'*) Top views of levelling plate
d) to *f*) Current flowing at an angle to the ship; bubble displaced at the same angle to the plate, and determined by reference to ship's heading

The bubbles are, therefore, no longer central. Now the position of the bubble can be determined with reference to the ship's heading by means of lines etched on the glass covers of the levels (Figure 81e); knowing the ship's heading, a true current bearing can be obtained. To get a measure of the current velocity it is necessary to know the relation between the tilt of the plate and the current strength and, although these meters have been calibrated indoors, they are better calibrated at sea for each depth against a standard direct-reading current meter. Once calibrated, the instrument is simple to use and requires very little maintenance. Usually a series of readings are taken at set times by a relatively untrained observer.

The drift indicator of Carruthers is a robust instrument, easily maintained and is capable of being worked by non-scientific personnel and of giving continuous records over long periods. It is a modification of the Ekman meter. A set of cups made from thick copper sheet and protected by a guard of stout wires is mounted on a spindle and this rotates under the action of a current (Figures 82 and 83). The recording mechanism is enclosed in a strong box, fitted outside with a projecting vane that orientates the instrument in the current. Internally a cross-bar carries a large lidded hopper containing a supply of phosphor-bronze balls. In the centre is a large crown wheel turned through gears by the rotating propeller and making one revolution for each hundred cup revolutions. A brass sphere with a cut-out slot rotates with the crown wheel and fits hard up against the hopper exit, so that only when the slit comes opposite the hopper are balls allowed to drop into the three holes (in a similar manner as in the Ekman current meter) (Figure 83c). Underneath is mounted the compass box and into this balls pass via a sloping runway, to be distributed by the compass needle trough into a lower collecting box. The number of rotations gives the average current speed, and by counting the number of balls in the various compartments the proportion of the time the current has been flowing in the different directions can be calculated.

The drift indicator is fairly expensive and, since direction is determined magnetically, for readings near the surface it can only be used from wooden ships. For these reasons Carruthers developed an even simpler device—the vertical log—a very cheap, robust instrument (Figure 84). The submerged part consists of a system of strong iron cups with heavy weights suspended below. The cup arms are welded to a steel bar and the top of the rod of the cup system has an eye from which a rope

Figure 82

The Carruthers drift indicator

(*Courtesy of H.M.S.O., London*)

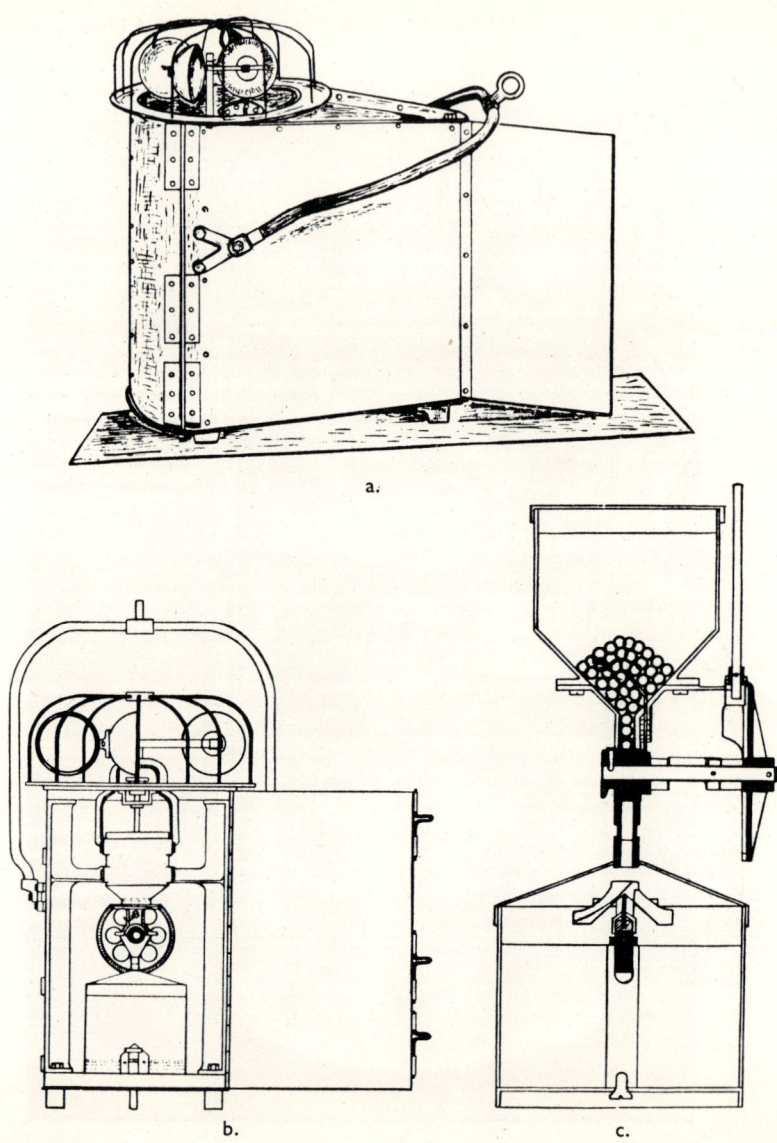

Figure 83   The Carruthers drift indicator
- *a*) External view
- *b*) Section
- *c*) Enlarged section of 'hopper' feeding balls on to magnet through hole in axle of wheel turned by revolving caps

(J. N. Carruthers: *a*) *and b*) '*Fishery Investigations*', 1928; *c*) '*Journal du Conseil*', 1926; *Redrawn from original*)

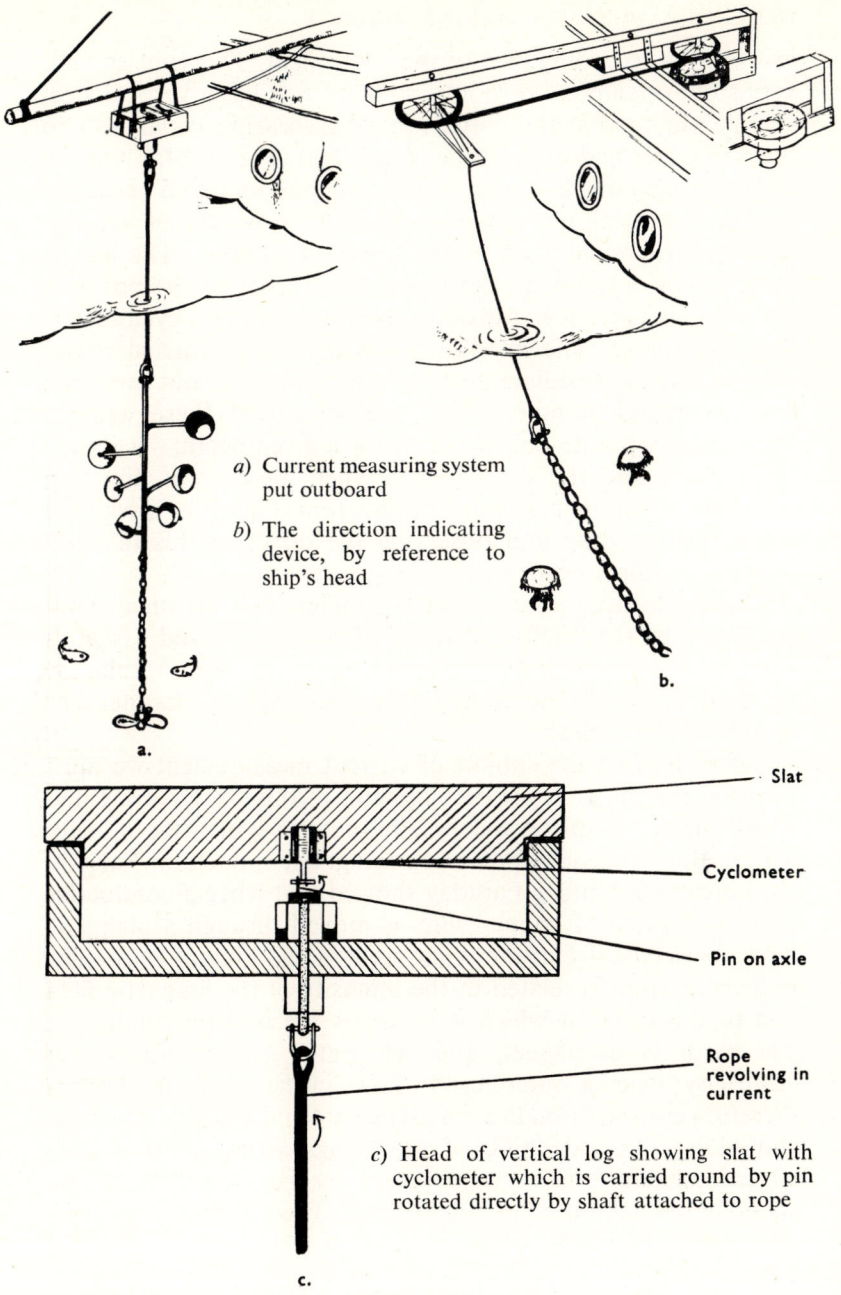

Figure 84   The Carruthers vertical log
(*J. N. Carruthers, 'Rapports et Procès-Verbaux'*, 1937; *Redrawn from original*)

leads to the recording part of the instrument. This consists of a very stout open box—a foot or so square. It has a vertical steel spindle mounted at the bottom in a ball-bearing collar and to this the upper end of the rope is attached (Figure 84c). On the top of the spindle is secured a large nut and mounted vertically in this is a steel pin about an inch long which, of course, describes a circular path as the spindle is rotated. The box is hung out on a beam. In the fore and aft sides of the box is a slit and this takes a hardwood slat which carries a cyclometer-type of counter. The handle of this counter is carried round without any intermediate gears by the pin on the nut mounted on the central spindle. There are several of these wooden slats and a separate one is put in for a given octant—a record being made and the slat changed when the current direction changes into another octant. The direction is taken by streaming out a floating drag and observing the direction this takes in relation to ship's head.

These robust continuous-current meters of Carruthers have been used by the English Ministry of Agriculture and Fisheries over long periods and have given some valuable results on net current speed and direction in the English Channel and Southern North Sea.

Before leaving the subject of current measurement we must mention the most recent development in this work—a method which though simple in theory is in practice somewhat difficult. The method depends upon electromagnetic induction. Early in the nineteenth century Faraday showed that when a conductor, such as a piece of copper wire, is moved through a magnetic field an electrical current is induced in the moving wire. The induced current is related to the intensity of the magnetic field and to the speed at which it is cut by the moving conductor. The earth is a magnet, and sea water is a conductor of electricity; when a water current flows in the sea a conductor is therefore cutting through a magnetic field and an electric current should flow. The possibility of measuring water currents by their electromagnetic effect was foreseen by Faraday himself; he stated in his Bakerian Lecture of 1832; 'Theoretically, it seems a necessary consequence [of the laws of electromagnetic induction] that where water is flowing, there electric currents should be formed: thus if a line be imagined passing from Dover to Calais through the sea and returning through the land beneath the water to Dover, it traces out a circuit of conducting matter one part of which, when the water moves up or down the channel, is cutting the magnetic curves of the earth, whilst the other is relatively at rest.' Faraday himself suspended two copper plates in a tideway, but with the instruments then available could not detect any potential difference other than that arising from chemical polarization at the electrodes. However, the effect was soon noted repeatedly on the very long ranges provided by broken submarine cables and quantitative and systematic experiments by Marc Dechevrens in the Channel Islands and by Young, Gerrard and Jevons in Dartmouth harbour clearly showed that tidal currents could induce electrical currents whose form and intensity agreed with that to be expected from the tidal streams.

It is to Dr von Arx of the Oceanographic Institute, Woods Hole, that we owe a further development of this method of measuring currents, namely, from a moving ship by means of what he has called the geomagnetic electrokinetograph. The undisturbed magnetic field of the earth is used as a frame of reference for the observations; it can be considered virtually constant under all conditions (even during magnetic storms).

The surface electrokinetograph consists of a pair of carefully selected silver-silver chloride electrodes mounted some metres apart on a two-conductor cable and streamed out from a boom over the side of the ship, the cable being long enough to take these electrodes away and astern from its influence (Figure 85). The electrodes are connected to a recording potentiometer (Figure 86). With this equipment and the ship's compass, observations are made under way of the direction of motion and the potential difference between the electrodes. These potential differences are due to water motion at right angles to the course and are related to the drift of the ship and the set of the electrodes. The potential difference changes sign when the currents set the ship to port or starboard. Although the principle is simple, the underlying detailed computations are

Figure 85
The geomagnetic electrokinetograph boom and cable towing astern of R.V. *Albatross III* in the North Atlantic. Surface currents were observed continuously along a course from Newfoundland to Ireland on this particular cruise
(*Courtesy of Mr Jan Hahn*)

SOME PROPERTIES OF THE WATER ITSELF

difficult and we must be content to state that the magnitude of the potential difference depends on the rate of current drift at right angles to the course, the length of water between the electrodes, the local strength of the vertical component of the earth's magnetic field, and, to some extent, the vertical distribution of water velocities in the vicinity. By measuring the potential differences on two courses nearly at right angles, the drift or component velocities in these two directions are obtained. The vector sum (or resultant) of these velocities gives the surface-water current vector for the locality. The signal strength is related to flow and total depth of water, and since in the open sea the flow is generally greatest at the surface with a vast body of highly conducting relatively quiet water below, these are the best conditions. Inshore currents (principally tidal) may extend almost uniformly from top to bottom, so that a quiet conducting layer is virtually absent and the circuit is completed mainly through the higher resistance of the sea bed.

In practice, a course of about 500 metres has been run, a limitation imposed

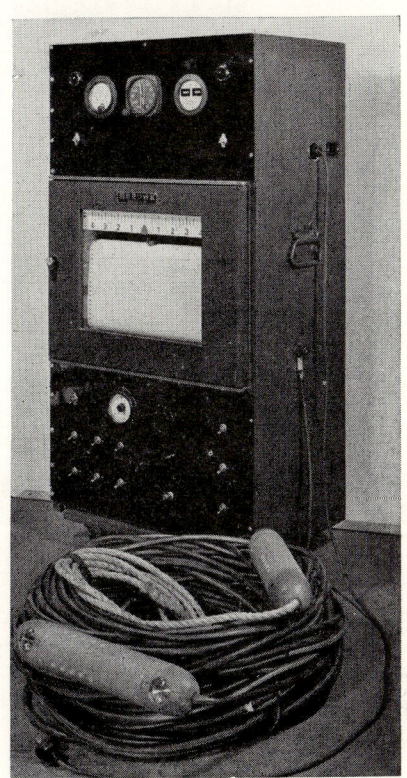

Figure 86
Deck unit, cable and electrodes of geomagnetic electrokinetograph, Model II. Used for five years on U.S.C.G. Cutter *Evergreen* on the International Ice Patrol, Oceanographic Surveys. Surface currents measured with this device every half hour under way between hydrographic stations provide data which replace simple interpolation between known points in the dynamic topography

(*Courtesy of Dr W. S. von Arx*)

155

OCEANOGRAPHY AND MARINE BIOLOGY

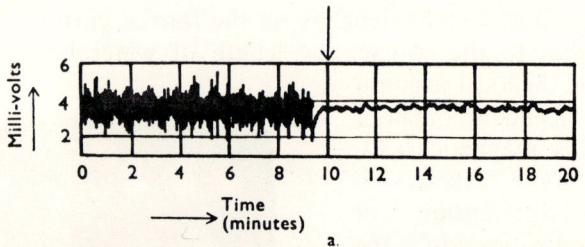

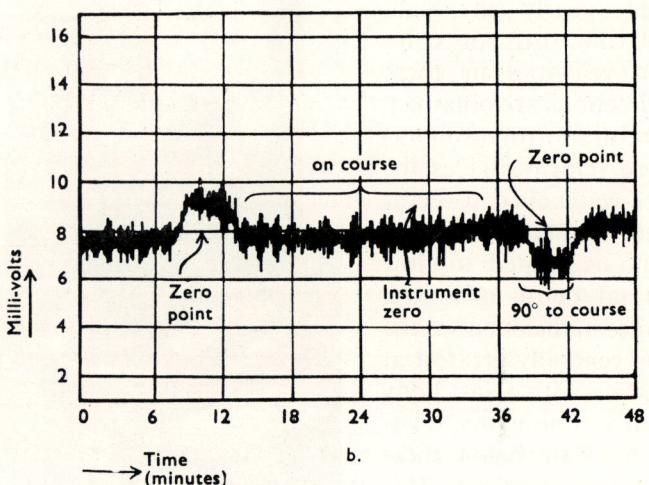

Figure 87

Recordings from a geomagnetic electrokinetograph

a) Example of a record: at the point marked the signals from the waves have been partially suppressed so that the main trace is clearer, but the small potentials from the waves are still present from which wave periods can be determined

b) Complete current record: long segment on course, short segment at right angles (90°) to course

(*Dr W. S. von Arx*, '*Papers in Physical Oceanography and Meteorology, Woods Hole*', 1950; *Modified from original*)

SOME PROPERTIES OF THE WATER ITSELF

partly by the apparatus and partly by the need to 'average out' the turbulent irregularities of motion in the sea itself. The type of trace obtained is seen in Figure 87. Here the long segment of the trace represents the component of water motion normal to the course and the shorter segments that at 90° to the principal course. It will be noted that there is a lag between signal and course change; this is directly proportional to the length of cable astern and inversely proportional to the speed of the ship. This record provides two estimates of true current, the intervening normal component to flow and also, from the width of the traces, data on wave periods; when the width is suppressed, the record is as shown in Figure 87b.

Reliable current-measuring techniques that detect identical aspects of flow are so few that it is difficult to devise conclusive experiments to demonstrate the validity of such a new method. It has been tested against propeller-type current meters suspended from an anchored ship, drift of current poles and dyes, predicted currents from various tables, and drift calculated from dynamic topography when steaming through a major current.

Since this instrument can be used with the ship under way, a large area may be rapidly surveyed and a synoptic picture of the current systems obtained. We may illustrate this from work that has been done in the vicinity of the well-known Gulf Stream. It will be remembered (pp. 110 and 127) that as one passes across the Gulf Stream there is a fairly sharp temperature-salinity boundary between Gulf Stream water and the so-called Slope water lying over the Continental Shelf. Figure 88 shows the surface velocity across this boundary at the continental edge of the Gulf Stream taken during a cruise towards Bermuda from Cape Cod and Cape May. This figure shows the existence of an entrainment current (that is, water dragged along) of Slope water on the continental side of the Gulf Stream. The entrainment current begins roughly 20 nautical miles from the boundary, and its velocity in general increases towards the Gulf Stream. The thermohaline structure shows (p. 110) that 'pools' of Gulf Stream water are detached from the main stream and the velocity structure suggests that the 'pools' have an eddy character, rather like a miniature whirlpool, the size of these eddies ranging from 3 to 10 nautical miles in diameter. Speeds as high as 3 knots have been observed near their

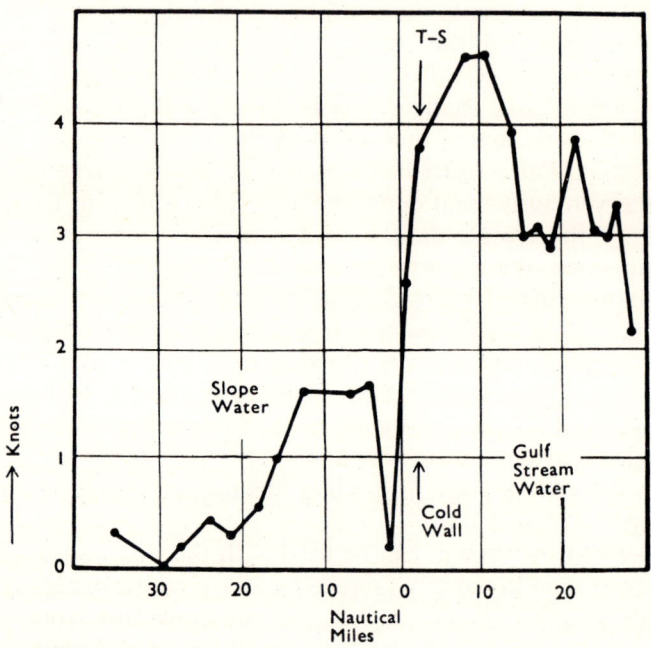

Figure 88
A velocity section along lines at right angles to the cold wall showing entrainment of Slope water by Gulf Stream and the abrupt transition in velocity at the temperature-salinity (T-S) boundary separating the two water masses. (See also Figure 57, p. 111)
(Dr W. S. von Arx, 'Papers in Physical Oceanography and Meteorology, Woods Hole', 1950; Modified from original)

margins. The velocity changes observed during a traverse indicate that these eddies may be moving bodily along with the Gulf Stream at about half its maximum velocity. At the main boundary of the Gulf Stream it is not uncommon for the velocities to double in a distance of only 3 nautical miles, reaching almost 5 knots without appreciable change of direction. The existence of the Gulf Stream has been known for many years, but only by means of these rapid recording instruments has its detailed structure been recently revealed.

4

PHOTOGRAPHY AND TELEVISION

AERIAL PHOTOGRAPHY

OUR primary concern throughout this book has been with problems involved in making observations and measurements within the marine environment: attention has, therefore, been focused particularly on methods by which a great variety of instruments can' be made to function properly in the sea. This section may, then, appear somewhat anomalous since, as the title indicates, it concerns a technique where the instruments are quite clearly remote from the sea. Although aerial photographs have sometimes been used to illustrate various aspects of marine biological studies—in particular those of the intertidal zone—comparatively recent developments have shown that the potentialities of this method are greater than had been realized.

The measurement of beach profiles

The necessity for widespread amphibious operations in World War II lead to an urgent demand by naval and military authorities for detailed information regarding beaches in various parts of the world; not only were pictorial records of many shorelines wanted but also estimates of beach slope and configuration, and their changes with weather over long and short periods.

Considered simply as records, aerial photographs of beach and shore-line topography have all the advantages of terrestrial surveys by aeroplane. Large areas may be covered rapidly, repeated excursions may be taken to determine any changes and, although the detail obtained is less than that given by conventional methods, aerial techniques may be far less expensive. Figure 89 shows a particularly interesting photograph taken from American work on the transformation of waves in inshore waters. When waves arrive in shallow water their form is dependent to a large extent upon the underlying beach topography and in this figure we see striking changes of wave pattern as the shore is approached. In this region a submerged canyon runs out at right angles to the shore and across the beach into the sea. The much greater depth of water over the canyon results in the waves being bent sharply forward, so that they are higher just outside the canyon than immediately over it. At the same time, directly opposite the mouth of the canyon, the surf zone is seen to be narrower than on other parts of the shore.

Careful examination of aerial photographs over shallow water reveals considerable underwater detail—the type of bottom, rock or sand, the presence of sub-littoral weed, and the changing pattern of sedimentation with the production of sand bars, can all be detected. Even the depth of water and hence the beach profile can be determined by aerial photography using an ingenious method worked out during World War II by Major J. Grange Moore of the British Army Photographic Research Unit.

Consider an aerial photograph taken over a uniform sandy beach. In a print, the sea bed will obviously appear clearer the shallower the water. Along a line extending at right angles to the shore into deeper water there will, therefore, be a gradual decrease in optical density (that is blackness) of the negative and this is equivalent to an *increase* in blackness of the print. The differences depend upon the change of depth of water along the line. A plot of the density of the negative against distance along any line is called the brightness profile. The light reaching a camera at any point vertically above a water-covered beach is made up of a number of components. First, there will

Figure 89 Aerial photograph near La Jolla, California
(From W. H. Munk and M. A. Taylor, 'Journal of Geology', 1947)

be light from all sources scattered by atmospheric haze between sea surface and camera; secondly, there will be a contribution from light not penetrating the sea but reflected upwards from the surface; thirdly, light which is scattered vertically upwards from the particles within the sea will also reach the camera; and fourthly, light will be reflected at the sea bed itself to emerge vertically from the sea surface. When we have constant atmospheric conditions and a fixed type of reflecting sea bottom the apparent brightness of the image appearing on an exposed photographic film will vary with only two factors, namely, the depth and clarity of the water. If the latter is known, then the depth profile may be determined by measuring the relative brightness of the image at different points along a line on a single photograph.

However, it may not always be possible to obtain a sample of the water and measure its clarity. An alternative and more convenient method may be used. This depends upon the fact that the clarity of the water, that is, its ability to transmit light (as measured by the so-called extinction coefficient), depends upon the wavelength of the light. Green light is transmitted better than red. If the relation between the extinction coefficients for two contrasted colours, say green and red, is known or can be calculated, then, using a special calculator, the depths may be determined from two aerial photographs taken simultaneously one through each coloured filter.

The aircraft, which need not be of a special type, carries a pair of cameras mounted vertically, one fitted with a red and the other with a green filter and simultaneous photographs are taken by synchronized shutters. The area covered by a photograph depends upon the height at which it is taken and the angle of view of the lens. During World War II much work was done from 5,000 ft., at which height a camera with an 8-in. lens gives a photograph 5 in. by 5 in. covering about a thousand yards square at a scale of 1/7,500. Although the area covered may be altered by changing the lens and the operational height, the ultimate scale should preferably lie between 1/7,000 to 1/10,000 (about 1/8,000 approaches the ideal). If the scale is too large, then, apart from giving insufficient cover, the photographs may show too much wave detail, while if the scale is too small errors arise from a less accurate location of points on the profile, whose brightness is being compared on the 'red' and 'green' negatives. All photographs must show a small amount of beach above the water and should be as free as possible from tilt, cloud shadows, and reflection from the water surface. It is possible to produce satisfactory pictures with broken cloud although this requires greater skill on the part of the pilot; if at all possible, the sky should be either cloud-free or fully covered with high cloud. To satisfy these requirements, as well as those considered below, the meteorological conditions have to be carefully considered beforehand. The time of day for taking the picture is also carefully selected; there is an optimum time for each lens angle. The sun should be as high as possible to give adequate illumination yet low enough to eliminate a tendency for specular reflection from small waves. The wind speed should be low, since apart from introducing navigational difficulties it both increases the amplitude of the waves, so tending to distort the brightness profile, and gives rise to local variations in the reflectivity of the sea surface. With high winds difficulties may even arise from the stirring up of bottom mud. Following exposure, the negatives are developed to full contrast under very careful control so that all parts of the film are under identical conditions of development. After being processed, although a knowledge of the absolute brightness is not required, the relative brightness in various parts of the negative must be very accurately determined. This is not so simple as it may appear. It is not sufficient just to measure, by means of a calibrated photometer, the amount of light that passes through various parts of the negative, because the relation between exposure time and density of the film after development is not linear. It is necessary to have a standard by which the relative brightness in different parts of the picture may be compared and an optical step wedge illuminated by a standard lamp is, therefore, exposed in each shot, the exposure being made synchronously with the picture by releasing the shutters together. A step-wedge consists of a series of small areas with increasing optical density; by calibrating it for each set of camera conditions, an arbitrary scale of brightness is obtained, and against this the relative densities over the negatives may be compared. A further and important correction must be made. When a picture is taken at a given shutter speed, the amateur taking his 'snaps' usually assumes, if indeed he thinks about it at all, that the whole of the negative is given the indicated exposure. This is not so; two factors vitiate this assumption. First, the amount of light which can pass through the lens is not the same at all points over the lens; in this respect each lens has its own performance. Secondly, the slot of a focal plane shutter does not pass across the whole negative at a uniform speed. A correction may be applied for both these 'errors' taken together by making control exposures under uniform lighting conditions, developing the film and comparing the optical density over what ought to be a perfectly uniform area. A correction template is then prepared on which the corrections to be applied in different parts of any subsequent negative are drawn in the form of correction contours. In reading a film, this template is placed over it and the necessary correction to be applied to any part of the film where the density is being determined can be read off directly. Figure 90 is a reproduction of three photographs by Major Grange Moore taken during his work.

Since sea water is more transparent to green than to red light, the 'green' prints show more underwater detail; under favourable conditions the bottom can be seen even at a depth of 50 ft. By contrast, 'red' prints emphasize the configuration of the shallows and enable qualitative estimates to be made of relative depths in very shallow water. The change in tone together with the sharpness with which rocks are outlined as the depth increases also gives some indication of the clarity of the water. In addition to the 'green' and 'red' shots, a picture is often taken on infra-red film. There is very little penetration of water by infra-red light and such prints are particularly useful for showing the exact position of the water line (when normal prints are made) or for assessing the depths in the extremely shallow water (when under-exposed prints are made).

If a quantitative estimate of the depth of water and the beach

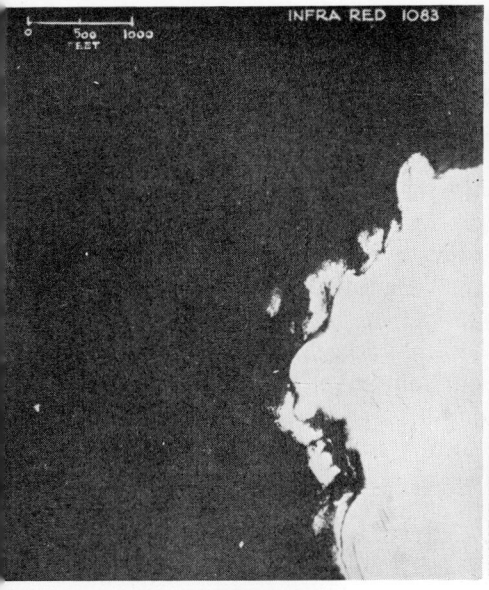
a)

b)

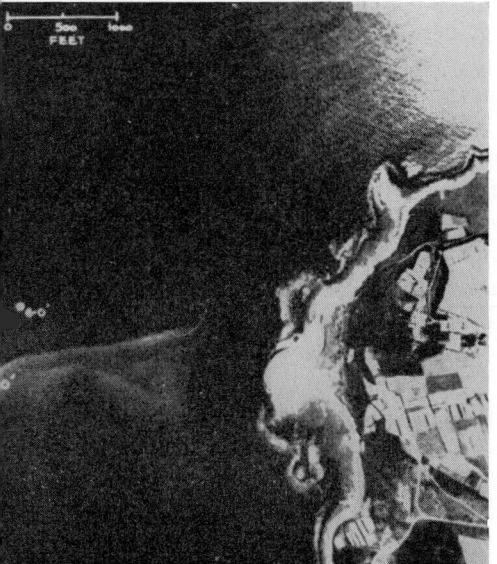
c)

Figure 90
Beach profiles determined by aerial photography. Bar Point, St. Mary's, Scilly Isles, taken at 10,000 ft. 1/250 sec.
a) *Infra red:* edge of coast shows clearly, completely submerged sandbar seen in other prints does not show
b) *Green:* great penetration, sandbar shows clearly
c) *Red:* less penetration, only shallow sand shows

(*Courtesy of Major J. Grange Moore;* '*Philosophical Transactions of the Royal Society*', 1948)

profile is required, then a brightness profile is set up; although reasonable estimates can be obtained from a carefully prepared print, it is better to work on the original negative. A line is chosen and this is carefully aligned either in relation to the water's edge (as obtained from the infra-red photograph) or from some selected point on the beach itself. The relative densities of the negative and of each step on the optical wedge (which, of course, appears on each negative) are then determined by an optical density comparator, an instrument that compares light passing through the film at any point and hence its density at that point. A comparison curve is plotted from the densitometer calibration data and the relative brightness along the selected line is then determined. When the brightness profiles for the 'red' and 'green' negatives are known, a special calculator is used to convert the results to actual depths which, over light-coloured beaches, may be determined with an accuracy of 10 per cent. over a range exceeding 20 feet.

Application to fishery research

A very different application of aerial photography and scouting —and one which may come to have considerable importance— is used in some fisheries work. Visual 'spotting' from the mast head has, of course, been practised for many years by some fishermen. Compared with this, however, an aeroplane can cover a far greater area more quickly and at any instant the observer has a larger field of view; further, sub-surface shoals of fish are much more readily seen from an aeroplane than from mast height. In practice, these advantages must be offset against the high capital cost of the aeroplane itself, the difficulty of handling it on board ship, particularly in bad weather, and the tendency for its rapid structural deterioration under working conditions at sea. The ideal aeroplane for the purpose does not exist, but it could certainly be designed. Helicopters could also be used.

Aerial scouting has indeed been practised in the Icelandic herring fisheries for some time—the aircraft being operated by the industry on a co-operative basis. An experienced skipper acts as observer; when herring shoals are located, the fleet is in-

PHOTOGRAPHY AND TELEVISION

formed by radio-telephone. This form of reconnaissance is said to be particularly valuable in the Icelandic herring fishery; traditionally the fleet tends to congregate in rather limited areas where the skippers, as a result of past experience, believe the herring will come to the surface and the searching efficiency of the fleet itself is thereby restricted. In addition, the echo-sounder is considered to be of limited application *in this particular* purse seine fishery, where the shoaling herring are caught near to the surface. In recent years aerial spotting has also been used in the tuna fishery of the east-central Pacific—the tuna boats are well adapted to carry planes because of their large free space aft over the bait tank. Shore-based planes have been used for inshore fisheries, as for example in the pilchard fishing off the Californian coast—in the years when this fishing was profitable.

Aerial photography and scouting have been used in preliminary exploratory work in Australia to give a broad picture of the distribution of pelagic fish shoals before beginning an expensive programme of detailed investigations. They have been used by Dr Eicher to estimate fish numbers. At the spawning season he took photographs (Figure 91), an example of which is given in Figure 92, of salmon in the rivers and lakes of western Alaska and, after magnification of the prints, an estimate of the number of fish could be made. For example, from this particular photograph he estimated that there were 1,600 salmon between the straight line and the mouth of the river. By repeating such surveys each year during the spawning season, changes in fish stocks were estimated.

Surveys have also been made from aeroplanes to follow the dispersion of waste discharged from barges and it is possible that variations in colour in the open ocean could be detected and their origin determined after directing a research vessel to the location.

UNDERWATER PHOTOGRAPHY

The development of aqua-lung equipment has led to a great increase in the number of people who take up diving and underwater photography either as a profession or as a hobby.

Figure 91

Aircraft of type most often used for visual surveys
(*Courtesy of Dr G. J. Eicher, Jr.*)

Frogmen have been used to take films for the purpose of biological research—for example, a film taken for the Scottish Home Department showing seine nets in action, and one for the Ministry of Agriculture, Fisheries and Food on the working of a trawl have given much new information; frogmen are being increasingly employed for special projects. Hand cameras for this type of work have been repeatedly described in recent books, and our attention will, therefore, be focused on remotely controlled or self-actuating underwater cameras which are now commonly used and which are playing an increasing part in marine biological research.

The scope of remotely controlled underwater cameras

We must first consider the scope of this technique and its relation to older and more conventional methods. Perhaps its most promising application lies in the observation of bottom-living plants and animals under natural conditions. It is possible with underwater cameras to examine organisms, both in relation to one another and to their environment, without the disturbance and consequent loss of detailed information inevitable when conventional sampling apparatus is used. It should also

PHOTOGRAPHY AND TELEVISION

be stressed that not only may qualitative observations be made but quantitative data on both the numbers and sizes of the animals can be obtained. In addition, observations may be

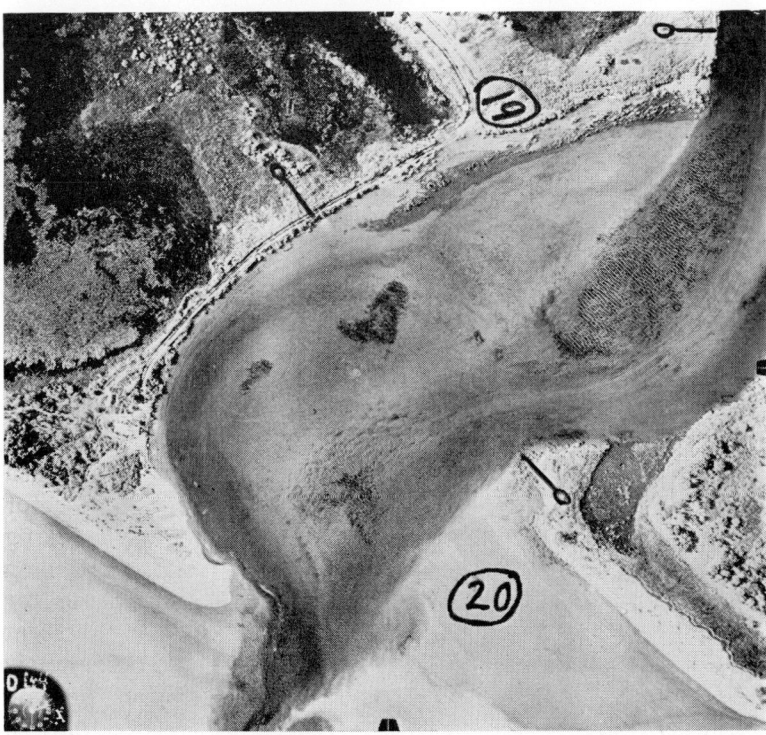

Figure 92
Photograph of red salmon at the mouth of Brooks River, Alaska. Taken at an elevation of 1,000 ft. An estimated 1,600 salmon are shown between the oblique line and the mouth of the river
(By G. J. Eicher, Jr.; Courtesy of U.S. Fish and Wildlife Service)

made on the physical character of the bottom and these supplement results obtained from examination and analysis of samples collected by conventional methods. It is also possible to use underwater cameras to record the results of experiments designed to test some particular hypothesis. For example, Professor Emery has proposed that turbidity currents, that is,

169

currents loaded with mud, could be investigated by placing a camera at the bottom of a slope and observing the effect of an anchor dragged across the upper part of that slope. The direct observation of the bottom also gives much information of

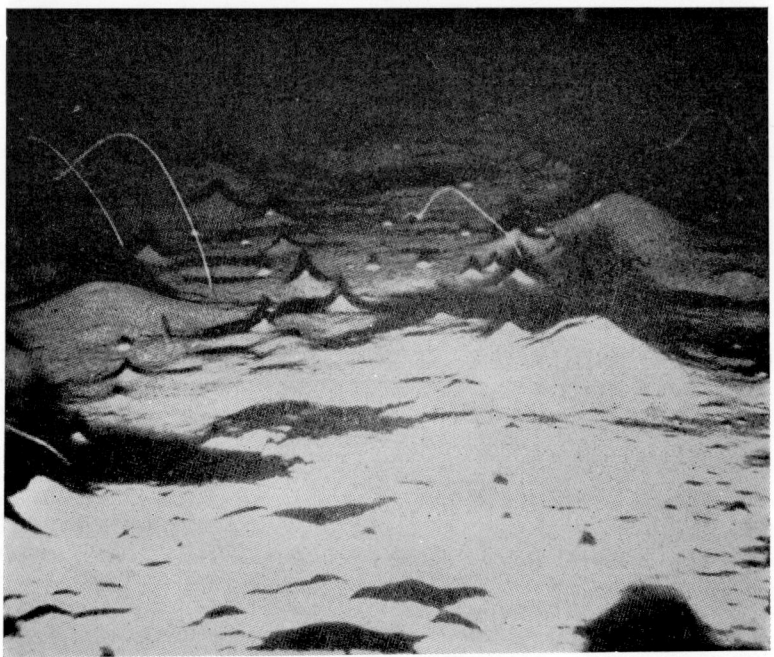

Figure 93
Underwater photograph taken by Professor Emery of the insular shelf east of Catalina Island (Lat. 33° 19′ N. Long. 118° 17′ W.) at a depth of 295 ft. This picture is of considerable geological interest; the sediment is a foraminiferous shell sand and the conical elevations are probably the result of worm activities

(*From K. O. Emery, 'The Scientific Monthly', 1952*)

geological interest (Figures 93 and 94); thus, Emery has pointed out that the surface appearance of a muddy bottom, as revealed by underwater cameras, contradicts the opinion generally held amongst geologists regarding the formation of evenly layered shales. The observations of track marks and surface features (Figure 95a, b and c) on muddy bottoms and the possibility of assigning them to known animals are also of importance to

PHOTOGRAPHY AND TELEVISION

geological studies (Figure 95). Many places such as the faces of steep underwater canyons are difficult to examine by any other means.

There is no doubt that underwater photography is a useful technique and that it can give information not otherwise obtain-

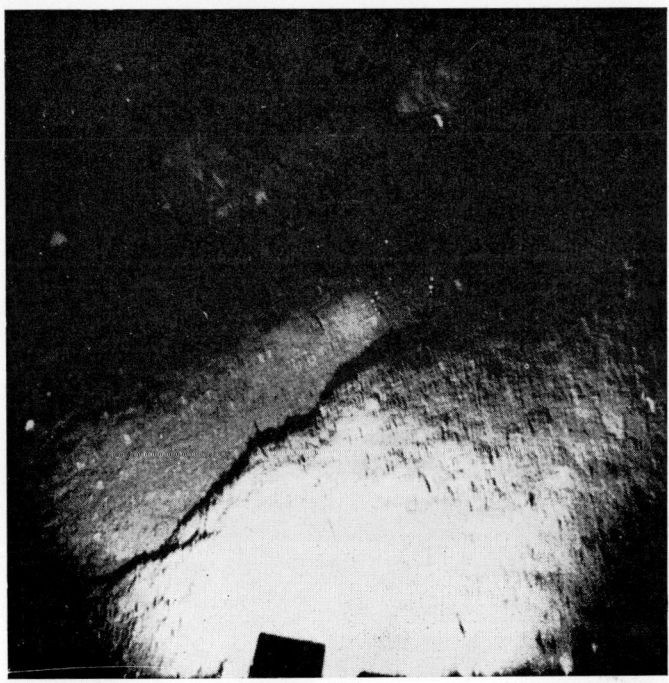

Figure 94

Outcrop of eocene; depth 1,000 metres. Photograph taken 90 miles south of Vinigard Light in 1947 and shows evidence of recent landslide. Lat. 39° 43' N. Long. 70° 48' W.

(*Original through courtesy of Dr B. C. Heezen*)

able, but like all other techniques it has its limitations. The subject, whether animal, plant or inorganic environment, is only 'inspected'; nothing is brought up for detailed examination or experiment in the laboratory. This is sometimes a severe limitation and in general ecological work the photographs should, when possible, be supplemented by conventional collecting and any other method of examination. The properties

171

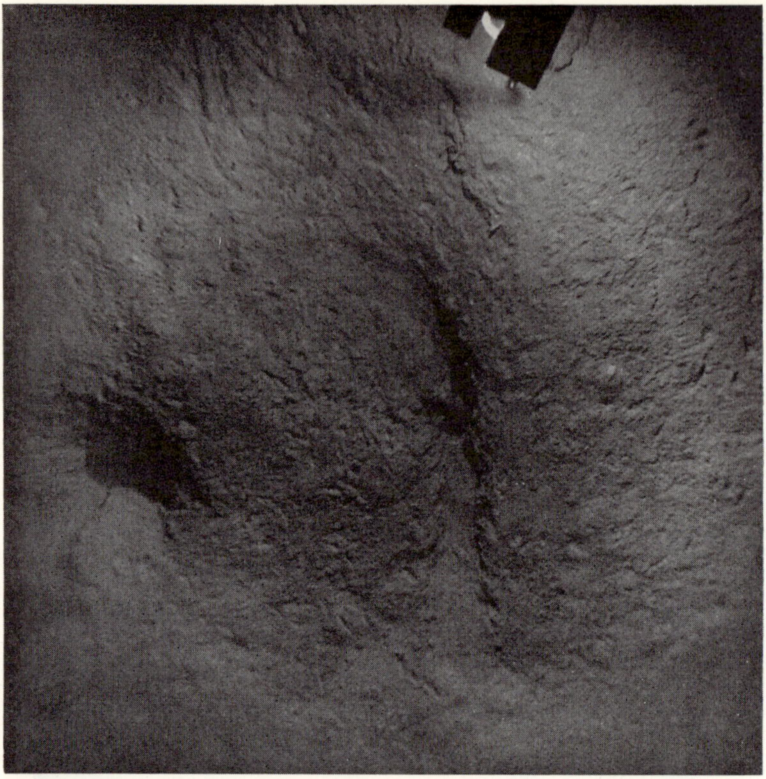

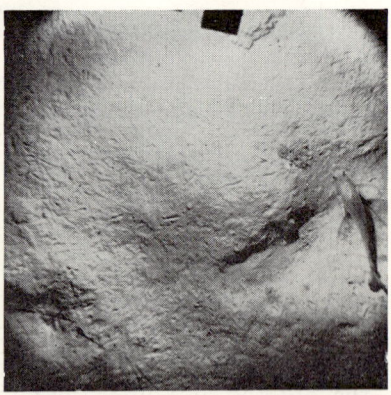

a) Mud-burrowing animals; depth 1,240 metres. Lat. 39° 29′ N. Long. 72° 13′ W. (*Courtesy of Dr. B. C. Heezen*)

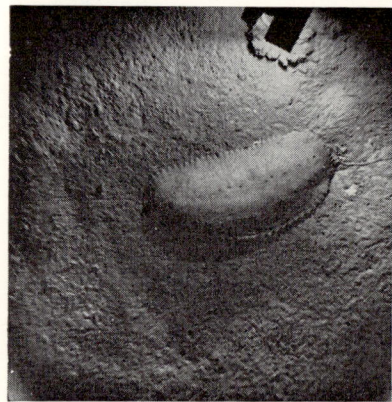 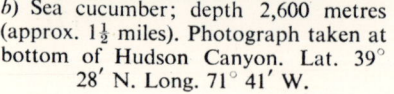

b) Sea cucumber; depth 2,600 metres (approx. 1½ miles). Photograph taken at bottom of Hudson Canyon. Lat. 39° 28′ N. Long. 71° 41′ W. (*Courtesy of Dr. B. C. Heezen*)

c) Deep water fish; depth 700 metres. Lat. 39° 55′ N. Long. 70° 51′ N. (*From J. Northrop, 'Bulletin of the Geological Society of America'*, 1951)

Figure 95

PHOTOGRAPHY AND TELEVISION

of the water itself impose the severest limitation; when the light transmission falls below about 80 per cent. per half metre, adequate results for research work become virtually impossible and the method can therefore only be of restricted application in some coastal, estuarine and polluted waters. Further, except in shallow water, when satisfactory results can be obtained by natural light, it is essential to provide artificial illumination of the underwater subject and this may affect the behaviour of any animals being examined. This must be constantly borne in mind when interpreting the results.

The problem of lighting

The question of light and lighting is of paramount importance to all direct or indirect underwater observations and we may consider it in some detail. When light passes through water some is lost by absorption but, except as this may influence the quality (spectral composition) of the light, such losses are of little importance; even if serious, they could be made good by the use of more powerful lamps (although the effects of this on living animals might then be more troublesome), longer exposures, and faster lenses. More serious is the effect of scattering, which produces a 'haze' so that a considerable amount of light not reflected from the subject enters the camera and destroys contrast.

The Admiralty Scientific Group have made a number of studies on this subject, particularly in relation to camera work by frogmen operating under natural illumination. They set up a large underwater step-wedge (see p. 163) and then by taking a series of pictures under varying conditions while approaching it, the effect on picture quality of water clarity, object tone, altitude of the sun, and working depth, were all investigated. The most important factor is clarity of the water; in the clearest water, with transmission greater than 90 per cent. per half metre, objects are recognizable at 10 metres. Below transmission values of 80 per cent. per half metre the working range drops sharply, becoming limited to a few metres, and there is a great reduction in picture quality. With regard to the tone of the object, the whites are always brighter than the sea background and the blacks always darker, whilst fairly dark greys match the sea in brightness even at close range. Under direct sunlight, greater ranges are obtained with white than with black objects, but the visibility of the latter is improved by looking towards the sun, when all parts of the target become darker than the surroundings, the black tones becoming more, and the white tones less visible. Using natural light, however, there is little that can be done to ensure good picture quality except to choose the right time and place to take the picture. These workers found there was no advantage in using a polarizing filter, although an attempt was made to detect a critical angle; a light haze filter was found to be of some benefit in clear water.

With artificial light, we must consider the scattering of direct light from the outgoing beam and that scattered on its way back to the camera from the illuminated subject. Scattering of the outgoing light can be reduced by illuminating as little as possible of the water between camera and subject, and this involves putting the lights well away from the former and near to the latter; little can be done to reduce scattering of the return light other than to minimize the distance between subject and camera. Except for short viewing distances, the best arrangement leads to a cumbersome lighting rig; a compromise must therefore be struck between the quality of picture that is acceptable (in relation to the distance one wishes to view and the particular problem in hand) and the cumbersome character of the equipment which can be worked under any given set of conditions.

Apart from the question of haze, there is the problem of the correct type and form of lighting. Under uniform illumination a picture depends upon the contrast inherent in the optical properties of the various parts making up the scene; if only such contrast is utilized, the picture is 'flat' and often difficult to interpret. To recognize solid shapes both adequately and easily it is necessary to have shadows and modelling effects whilst still ensuring that the former are not so dense as to reduce all contrast in the darker parts of the picture. From this point of view uni-directional lighting, particularly from directly above the subject, is inadequate; both top and side lighting are necessary, as is well known from studio work. Again it is a question of compromise. A multi-lighting system with the individual units under separate control is cumbersome, but to ensure good picture quality as much flexibility of this kind as can be accepted is desirable. Commonly, a single light placed somewhat to the side is used, but it is then found that in general the foreground is over-illuminated and the background under-illuminated.

Various types of artificial light have been used, both continuous source such as a tungsten filament lamp, or discontinuous, such as flash bulb and electronic flash discharge. When working to moderate depths power losses of the cable are insignificant and handling of the latter is relatively simple; there is then much to be said for continuous-source lamps supplied by power from the ship's mains. Such lamps are cheap and easily replaced, no high voltages are involved, and no separate case (or an enlarged camera case) is required to house a separate lighting unit. There is some difference of opinion as to whether anything is gained by the use of mercury or sodium discharge lamps in comparison with the simple tungsten filament variety. The latter are simpler and cheaper because they require neither a warming-up period nor accessories such as chokes, transformers or ballast resistances.

The selection of equipment

For deep-water work, power losses and the difficulties involved in handling long cables become overriding considerations and a self-contained lighting unit with either flash bulbs or electronic flash equipment is almost invariably used. The use of flash bulbs severely limits the number of pictures that can be taken at one lowering and they have been largely superseded by electronic flash equipment. It might be added that with flash photography, if the number of repetitions in one place is limited, abnormal animal behaviour due to the effect of light is less likely to occur.

For shallow and moderately deep water there is much to be said for a camera either remotely controlled from the deck or actuated when it contacts the bottom, but in both cases with the necessary power supplied by cable from the ship. Problems of spillage of batteries and their accommodation in a watertight case are thereby eliminated, and the apparatus can be quite small and therefore easy to make and handle.

Shallow-water work

When working to moderate depths it is only necessary to put a

Figure 96

Two views of underwater camera for moderate depths, developed by Mr R. E. Craig of the Scottish Home Department. Above, interior view. Below is seen its window mounted eccentrically on the case and the reflector for the electronic flash lamp

(*Courtesy of Mr R. E. Craig*)

camera in a suitable watertight case and to provide a lighting system giving a form easy to handle. It is convenient to mount a Robot-type camera in a watertight case, usually fastened to a pole (Figure 99). A single light, a battery of floodlights or electronic flash equipment may be attached to this pole. For pictures

Figure 97
Photograph of a shell-covered bottom in Loch Creran by Mr Craig. The triggering weight is seen in the top centre. Ophiuroids, starfish and algae prominent
(*Courtesy of Mr R. E. Craig*)

of the bottom the camera and flash-light are actuated by a foot switch mounted at the bottom of the pole, or by a weight attached to the lever of a mercury tipping switch. When the foot or weight hits bottom, the mercury switch is tipped and this then activates the solenoid in the camera case and when electronic flash is used it is flashed; on completion of the circuit the shutter is released and the film automatically wound on. At the same time a 'camera' ammeter on deck gives a reading, a frame counter is actuated and a buzzer may be arranged to alert the crew.

The apparatus is lowered on a wire warp and speed reduced when nearing bottom; if continuous-source lighting is used, the lamps are then switched on. The camera is gently lowered on to the bottom, the foot plate or weight contacts and a picture is taken. For detailed survey of an area the apparatus is then lifted off the bottom and the ship allowed to drift; the process may be repeated as required, so that a series of pictures can be taken on the line of the ship's drift which is subsequently plotted on a chart. For recognition and counting, the negatives are examined under a binocular microscope or enlargements are made (Figure 97).

Deep-water equipment

In deep-water work two types of apparatus must be considered, namely free-floating and captive. In the former, the apparatus is made lighter than water by the addition of floats and taken down under ballast; on contact with bottom a picture is taken, the ballast is released and the camera comes to the surface. In the second type, the apparatus is lowered on a cable as with shallow-water gear. The advantage of the former is that no cables are required and no time is wasted in lowering—an important factor when depths of say several thousand metres are being worked; the camera can be put overboard and the ship is then free to carry on with other work until it returns to the surface. Picture quality is good because the exposure is made while the apparatus remains motionless prior to release of ballast; there is no strain, as with a suspended type. On the other hand, only a single picture of the bottom is taken at one lowering although it is possible to arrange for the apparatus to take pictures on the way down. The ship will drift while the apparatus is out and marker buoys must be put out to facilitate return to the shooting position.

A free-floating camera which takes a single picture at each lowering has been described by Ewing, Vine and Worzel. The camera case is mounted on a pole below the lighting unit. Two types of ballast release have been used; in one, contact with the bottom actuates a trigger that releases a clockwork mechanism in the camera and after two pictures have been taken the ballast is let go by an electromagnetic mechanism. In an alternative and less precise method, a bag of salt is attached to the gear and, when the former has dissolved, the containing bag is pulled through a hole, thus dropping the ballast; protection from solution during descent is effected by a shield plate that falls away when bottom is reached. For depths not exceeding 100 fathoms glass or metal floats of conventional design may be used, but for deep water a flexible fabric or neoprene bag containing petrol is convenient; such a liquid float readily adjusts its volume to pressure and temperature changes during descent.

The limitation of a single picture at each lowering is a very severe drawback with this type of camera and in recent years the suspended-cable type has been largely preferred. The free-floating variety may still be of use as an 'occasional' instrument, for example, to take a bottom photograph whilst other work is in progress at a station; in general, however, for an intensive bottom survey a large number of photographs over an area or on a transect is required.

Two deep-sea cameras of modern design will be described. The USNEL Deep Sea Camera (Figures 98 and 99) was designed at the U.S. Navy Electronics Laboratory and has been largely used by the personnel of the Scripps Institution, California. It is designed for depths up to 1,400 fathoms and takes 40 exposures at a single lowering, each picture covering 30 sq. ft. The camera is a 35 mm. robot with a spring operated film advance and a 37·5 mm. f2·8 Schreider-Kreuznash lens. 100 A.S.A. film is used, working at f6·3-f8·0 and a shutter speed of $\frac{1}{25}$ second. The light source is a commercial electronic flash with a repeating condenser discharge; the unit automatically recharges in about 7 seconds and gives up to 100 flashes before it is necessary to recharge the batteries. The camera and lighting unit are housed in separate cases of special construction (Figures 98 and 99), the camera case being so designed that its cover needs less than two complete turns for the proper seating; it can be quickly opened and closed. Both units are mounted on a vertical frame (Figure 99a), the whole apparatus weighing 400 lb. in air and 300 lb. in water. The camera mechanism is actuated on

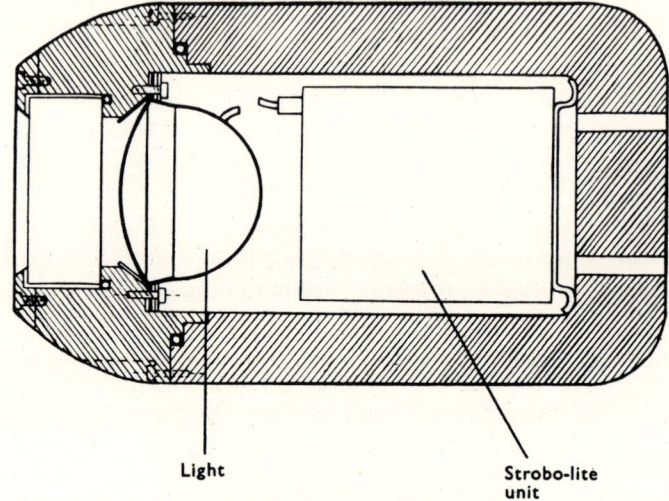

a) Section through lighting unit

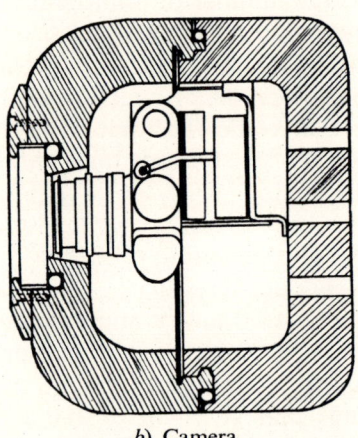

b) Camera

Figure 98
The USNEL Deep Sea Camera and lighting unit

touching bottom by the tripping foot and mercury switch as already described for shallow-water cameras.

The Deep Sea Benthograph (Figure 100) developed by the Hancock Foundation has both the lighting unit and camera

PHOTOGRAPHY AND TELEVISION

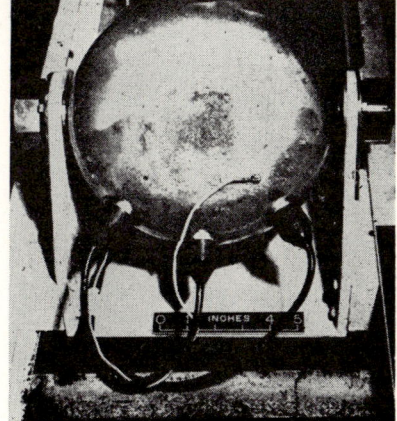

Figure 99 *a*) USNEL camera being put over
b) *c*) Camera

contained in a single case which can rest on the bottom; there is no subsidiary framework. The apparatus is larger than that just described, since it was designed so that it could easily take other deep-sea equipment. The watertight housing consists

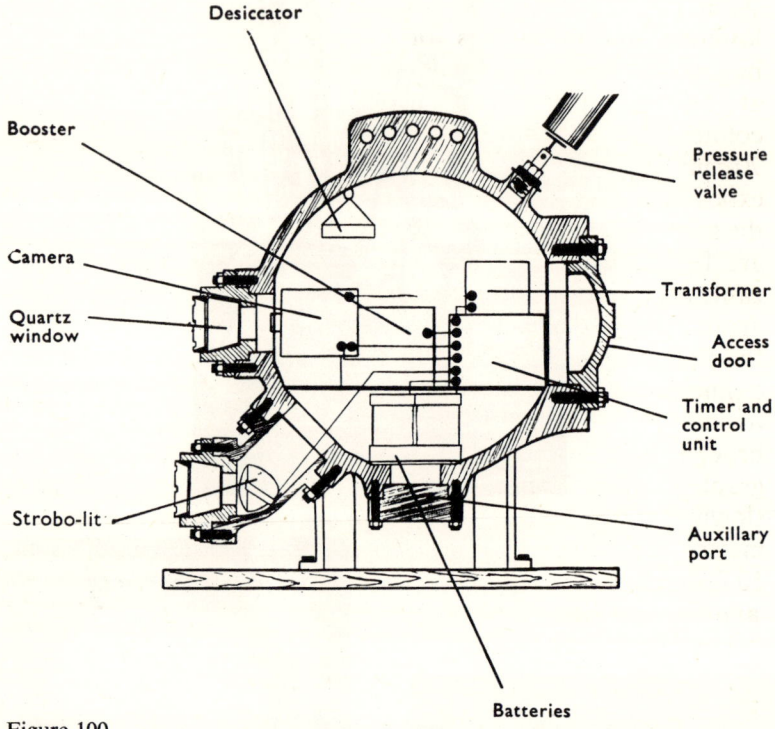

Figure 100
The Benthograph
(K. O. Emery, 'The Scientific Monthly', 1952; Redrawn from original)

of a hollow steel ball 39 in. in diameter with $1\frac{1}{2}$ in. walls, four legs at the bottom and five holes for cable attachment, all forming an integral part of the casting. The sphere has five openings, the largest of which is a 15-in. diameter access door closed by a cast steel plate bolted in place and sealed by two neoprene O-rings. Three other similar ports are available with windows 3 in. thick and 6 in. in diameter, and hollow extension arms can be inserted in these ports and the windows then moved to their outer ends. The fifth opening is a pressure-release valve. The weight fully equipped is 3,300 lb. in air and 2,150 lb. in water—far greater than the USNEL apparatus. The winch by which the equipment has been handled takes 12,000 ft. of $\frac{7}{16}$ in. non-spin cable; about 10 minutes is required per 5,000 ft. lowering and 35 minutes for hauling the same distance. The best results have been obtained with exposures of $\frac{1}{50}$ second at f5·6 when using Agfa Plus X film or at f4·5 with Agfa colour films. When fully loaded, the camera holds 100 ft. of 70-mm. film and gives 25 exposures at a single lowering, the exposures being made at 30-second intervals. The large size of the sphere allows much bigger film to be used and the negatives are less 'grainy' than those from a smaller 35-mm. type of camera. Electronic flash-lighting with a G.E.C. flash-tube is again employed, the camera mechanism being synchronized to the flash. Instead of using a bottom release to start the camera, a time-delay mechanism, set for the depth at which it is intended to work, is used: with this arrangement no cables need be brought outside the pressure case. When in use, the benthograph is lowered rapidly until near the bottom and then more slowly: after touching bottom, as indicated by a cable-strain meter, it is raised to wash off mud, then lowered to within 10 ft. of the bottom and pictures taken as the lowering proceeds and the ship rides to the waves.

UNDERWATER TELEVISION

Underwater television is one of the latest techniques of marine biology. In 1947 a few isolated tests were made in America at the Cornell Aeronautical Laboratory, but the possibilities were not further explored. Independently in 1948 a programme was initiated at the Millport Marine Station and the possibilities of the technique from the marine biological point of view were demonstrated at the London Zoo. It is regrettable, in the light of subsequent events, that some British manufacturers could then see no future in underwater television. Whilst these developments were taking place and equipment was being prepared for marine biological work, underwater television was dramatically brought before the public. Soon after the tragic loss of H.M.S. *Affray* early in 1951, it became apparent that search for the wreck could be resolved into the problem of identification of the *Affray* from amongst many unknown wrecks which had been detected and positioned as a result of an intensive acoustic survey following the accident. Scientists of the Royal Naval Scientific Service, who were already thoroughly versed in underwater camera techniques, rapidly designed and built a strong underwater casing to hold a television camera supplied by the Marconi Company. The *Affray* was eventually recognized on June 14th, 1951, and underwater television had come to stay.

Since then, the Royal Navy has continued to use and develop the technique for its own rather specialized purposes and underwater television equipment may now be considered part of the standard gear of a really modern vessel fitted for deep-diving work such as H.M.S. *Reclaim*, on which the original Navy apparatus was installed.

Here we are concerned more with its application to marine research: as a result of the preliminary experiments already outlined above, His Majesty's Treasury in 1949, on the advice of the Development Commissioners' Advisory Committee on Fisheries Research, made a special grant from the Development Fund to the Scottish Marine Biological Association for the development of underwater television. The laboratory of this Association at Millport on the Clyde is, in many ways, particularly well suited for such development work; near at hand is deep sheltered water oceanic in character and, therefore, much less turbid than is usually found in coastal regions. Since this project was begun, developments of a similar character have taken place elsewhere—and in the field of freshwater biology the Canadians have used underwater television in work on their lakes.

The principles of the television camera

In principle, underwater television is simple; once the necessary electronic equipment became available it was a logical development of remotely controlled underwater photography. It is only necessary to enclose a television camera in a suitable watertight case—sufficiently strong to withstand the pressures at the maximum required working depth—to provide lights for illuminating the underwater scene, to bring up the signals generated, and re-constitute these into the original picture. The circuit is a closed one with no radio link. In order to understand its advantages, we must look a little more closely into technical details.

The physical requirements for the reproduction of a scene are, first, a method of focusing the scene on to a sensitive recording surface, and secondly a means of re-constituting the original. In the photographic camera the scene is focused on the sensitive surface by adjusting the distance of the lens from that surface and the amount of light passing through the lens is controlled by an iris diaphragm placed immediately behind the lens. All these adjustments are usually made mechanically. The film is the sensitive surface and its individual elements are the light-sensitive granules of silver salts (together with other substances), which undergo a chemical change in relation to the intensity of light falling on them; this change allows, through the action of developer and fixer, the tone values of the original scene to be made permanent in the finished negative. The tonal values are, of course, reversed in the negative; the exposure of this to sensitive printing paper and subsequent development gives the positive print in which the tonal values of the original scene are reproduced. The resolving power is determined by the size of the individual sensitive granules. The illusion of movement in ciné work is only obtained by throwing successive still pictures on a screen at such a rate that the eye, as a result of the persistence of vision, is unable to detect the discontinuities.

All these features are present in a television camera—but it has the great advantage over ciné work that the scene can be transmitted virtually instantaneously. Lenses similar to those used in photography focus the scene on a sensitive surface, the so-called mosaic of the television tube (see Figure 102a), but focus is effected by moving the sensitive surface in relation to the fixed lens. The light-sensitive surface is again a mosaic consisting of a large number of elements—but in this case the light does not produce a permanent chemical change in these elements; on each element there develops an electrical charge proportional to the light intensity. This phenomenon is termed the photoelectric effect and it depends upon the property possessed by certain substances of releasing electrons (either internally or externally) under the action of light. The saturated photoelectric emission of any picture element is proportional to the light falling upon it, that is, to the brightness of that picture element. In practice, these minute photoelectric elements are associated with a micro-condenser on which a charge is built up. It is, therefore, possible to reproduce the tonal values of all the picture elements of a scene in terms of electrical quantities of equivalent magnitude. The most that any television system can do is to reproduce the average brightness over each mosaic element, any gradation in detail within a single element being lost. The greater the number of picture elements and the greater the number of tone gradations that can be produced between black and white, the more perfect will be the reproduction obtained. As with other methods, the size of the individual picture elements determines the ultimate theoretical limit of the resolving power of the system—although in practice other factors limit the resolution before such a theoretical possibility is attained. The mosaic of the C.P.S. Emitron tube, for example, is made up of a large number of these photosensitive elements—about 2.5×10^6 elements in a mosaic area of 35×44 mm. In a 625-line picture of 4×3 format there are some 5×10^5 picture points, which gives about 5 mosaic elements to each one.

The C.P.S. Emitron sensitive mosaic—or target as it is often called—is prepared *in situ* that is, inside the evacuated camera tube. The most efficient photoelectric surface for the visible spectrum is a layer of antimony treated with caesium to form an alloy. The target, a sheet of transparent insulating dielectric such as mica, is 'backed' by a transparent metal layer (the signal plate) which faces the glass window through which the scene is focused on to the mosaic. The sensitivity of such a transparent antimony-caesium photo-surface is between 30 and 40 micro-

PHOTOGRAPHY AND TELEVISION

amperes per lumen (μA/lumen). After allowing for light losses at signal plate and dielectric, the overall net sensitivity is of the order of 10-17 μA/lumen.

Scanning and picture regeneration

To regenerate the visual picture, the scene present on the mosaic in terms of electrical charges must be converted back into light intensities. The basis of this conversion in modern television practice is the cathode ray tube, one type of which has already been described (see p. 82).

We have seen that in a television camera tube the scene is built up in terms of electrical charges on the elements of the tube mosaic; these must be transmitted to the receiver. It is not possible to transmit all these picture elements simultaneously and the picture is, therefore, dealt with in an agreed order, the process being known as scanning. This order is similar to that involved when reading a page of print—each element is taken from left to right in a line and each line is taken successively downwards. Scanning then begins again at line one (Figure 101).

The cathode ray tube constitutes an excellent scanner and so the mosaic is built into a tube rather like a cathode ray tube but with the mosaic replacing the fluorescent screen; an electron beam moved by deflecting coils scans the target. The extent of the deflection is proportional to the strength of the magnetic field at right angles to the axis of the tube, and therefore to the strength of the current in the deflecting coils. To scan one line it is necessary to deflect the beam to the left-hand edge of the rectangular area to be occupied by the picture. This current must then be reduced to zero and beyond to an equal strength in the reverse direction. Then it must be rapidly reversed to bring the spot to the start of the next line—to the left-hand edge again. At the same time it is deflected downwards to the beginning of the next line at its left-hand edge. Each complete picture is called a frame and the number of complete frames per second is the frame frequency. In practice, a system of so-called interlacing is used. First the odd lines are scanned and then the even lines. In this way, although the full detail of the picture is spread over $\frac{1}{25}$ second, the flicker frequency is only 50 per second (selected in relation to ordinary mains frequency) and this is too high to be noticeable even with a bright picture.

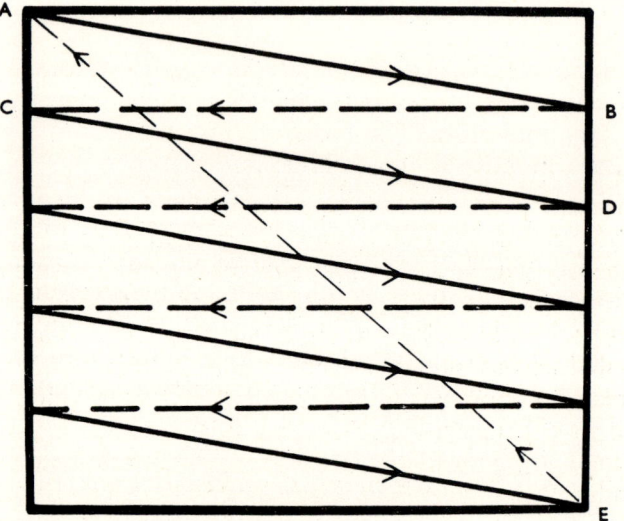

Figure 101

The scanning process

The electron beam scans line 1 across from A to B; it is then returned to the left-hand edge, the signal being suppressed while doing so; it then scans line 2 from C to D. At the end of the frame E it returns to line 1 at A during frame suppression

At the receiver the train of electrical impulses resulting from scanning of the mosaic is applied to the grid of an ordinary cathode ray tube with a fluorescent target so that the brightness of the receiver spot varies with that of the camera.[1] In order to keep transmitter and receiver in phase, there are synchronizing signals both at the end of each line and each frame, all of which are kept lower than 'black' level and therefore not seen on the viewing screen.

[1] In image orthicon cameras the system is somewhat different; the mosaic itself is not scanned but only a secondary target. Further, several stages of electron multiplication are interposed before the impulses are passed to the receiver.

PHOTOGRAPHY AND TELEVISION

The Millport equipment

We shall now describe the equipment which has been used especially for marine biological work in Britain and in which the C.P.S. Emitron system of E.M.I. Research Laboratories Ltd. is employed. In other underwater cameras image orthicon tubes have been used; both systems have their own special advantages, and neither would appear to be perfect for all types of work.

The camera tube with its scanning coils, the head amplifier which magnifies the original signals, together with the scanning unit, are mounted (via anti-vibration mountings) on a movable carriage, so that the scene can be focused on the mosaic by the lens (Figures 102b and c). The whole camera tube assembly is moved backwards and forwards in relation to the fixed lens by a magslip motor mounted under the carriage (Figure 103). This, as already mentioned, is in contrast to photography, where the sensitive surface is kept fixed and the lens moved to bring the scene in focus on the film or plate. The anti-vibration mounts are particularly essential on board ship; they reduce spurious signals due to any mechanical vibration of the valves. All the connections are taken to two plugs at the back of the camera. A revolving turret (Figure 102b) is mounted at the front of the camera and carries six lenses which are held by bayonet fittings and locking rings. A lens is selected by rotation of this turret, which is driven by a motor situated at the rear. The diaphragms of the lenses are controlled through gears by a motor on which the turret itself is mounted. All the lens diaphragms turn together so that when a lens change is made at any given aperture, then that aperture is maintained during and after the change.

The signals from the camera (housed, of course, in its cylindrical watertight case) as well as the supplies to it and the circuits of the various control mechanisms are connected to the equipment via a 32-core P.V.C. cable, which is 2 cm. diameter and has a double-walled outer covering both for added strength and as protection against rough handling.

The six camera-control units are of very compact design

OCEANOGRAPHY AND MARINE BIOLOGY

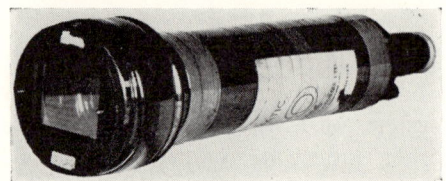

a) C.P.S. Emitron camera tube. The mosaic can be seen through the front window

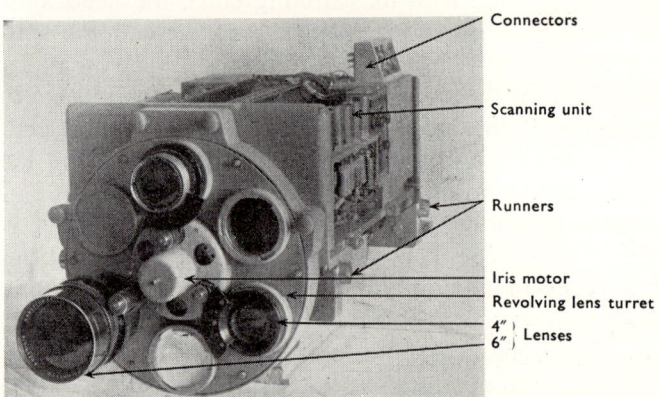

b) The camera. Removed from water-tight case

c) The camera. Side view with head amplifier removed

Figure 102

PHOTOGRAPHY AND TELEVISION

Figure 103   The camera. Underneath view

with carrying handles, and they are housed on board ship (Figure 104). The names of these units indicate their function, and we can only deal with them briefly. The voltage-control unit regulates the mains input and so supplies the equipment

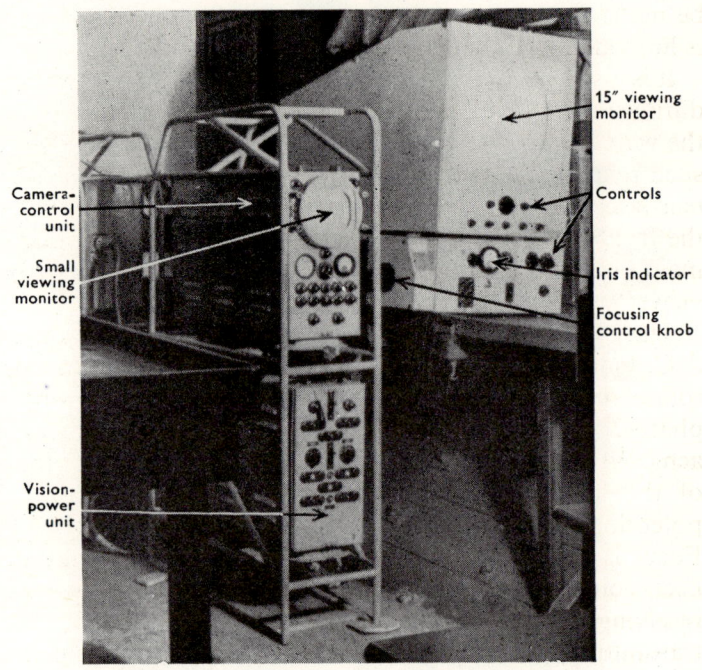

Figure 104
Inboard units—mounted in ship's laboratory

with stabilized alternating current at the appropriate voltage. The waveform generator produces the necessary pulses for the whole camera channel. The camera-control unit contains all the electronic control circuits for the camera. Power to the camera itself is supplied from a fourth unit, known as the vision-power unit.

This set-up seems very complicated but in practice, once the channel has been set up, only the camera-control unit requires adjustment—and this usually only to a limited extent. On this unit is a small 6-in. cathode ray tube used to check the picture quality during preliminary adjustments. It is from this small tube that film or ciné pictures are taken.

The various remote controls are mounted in the fifth unit—a junction and control box, and adjustments can very rapidly be made. A large 15-in. viewing monitor with its own controls is housed on the top of the junction box.

It is essential to be able to control certain camera mechanisms during viewing and since the camera may be 600 ft. under the water, this must be done by remote control. There are three such remote controls, namely, optical focus, lens aperture, and lens selection. The control knobs for all these are mounted on the front panel of the junction box (that is, directly under the main viewing monitor) where they are readily reached by the viewer.

The iris motor controlling the lens aperture is a split-phase A.C. induction type fitted with a reduction gear and this will rotate in either direction depending upon which winding is phase-displaced with respect to the incoming mains. Control is achieved by a pair of micro-switches mounted on the front panel of the junction box. By means of a calibrated logarithmic potentiometer coupled to the shaft, the iris setting is known. Focus is achieved by a 'magslip' control which, as has already been pointed out, moves the tube-carriage and not the lens. The receiving magslip is suitably geared to the carriage and the transmitting magslip with both coarse and fine controls is housed in a detachable unit on the junction box. A divided drum is fitted to the focus control, each section being calibrated in object distance for a given lens; indicator lamps illuminate the scale relating to the lens in use, the appropriate lamp being selected by the lens-change switch. In this way, with the control knob in any given position, the distance at which the lens is focused may be read. The turret in which the lenses are mounted is rotated by a motor similar to that used for the iris control and the circuit has been arranged so that the turret rotates from any given position in a direction which gives the shortest route to each lens.

The camera is housed in a watertight casing strong enough

to stand the water pressure at the deepest working level. A simple steel cylinder 18 in. diameter and 30 in. long with flanged ends has guide rails on which the camera slides into position, being held firmly by brackets and captive screws. Both ends of the case are domed bronze pressure castings seated on rubber gaskets and fastened by a ring of stainless-steel studs. The plate-glass window (10 in. diameter and $1\frac{1}{2}$ in. thick) in the front casting is mounted axially in alignment with the operational lens. It is seated on a rubber gasket and clamped in position by a bronze ring and pressure gasket. The rear end casting has a small service port some 10 in. in diameter secured by quick release studs so that minor adjustments can be quickly carried out without removing the heavy rear cover. Two bags of silica gel held at the sides of the camera unit keep the air inside the case dry and prevent condensation of water, which would fog the window and lenses. A simple leak detector is fitted inside the casing; this is made from two brass plates which are insulated from one another and separated by a small gap; if this air gap is bridged with sea water, a current passes between the plates and warning is given on a buzzer in the junction box. In this way a leak may be detected before any serious damage is done.

Eight eyebolts are welded on the case and take bridles for either vertical or horizontal suspension, the bridles being taken from the appropriate eyebolts to a single shackle. Eight symmetrically placed hollow lugs are also welded to the case and into these short pieces of scaffold pole can be fitted. It is from these that the lighting gantry is built up. The lighting distribution box is mounted on top of the case (Figure 105).

For experimental purposes a flexible lighting system is desirable, so that changes in the lighting set-up can be made and their effects observed. This has been achieved by building up a framework—or gantry—from standard scaffold tubing and swivel clamps (Figures 105 and 106). This framework projects from the camera casing and it is stayed from the short pieces of scaffold tubing attached to lugs on this case. Such a framework can be built up to various sizes, the angles of the outriggers always being set off to be rather greater than that of the widest angle

PHOTOGRAPHY AND TELEVISION

lens to be used. With the large frame, which may rest on the bottom, an area of some 50×35 in. can be viewed with more or less lateral lighting when using the widest angle lens (2½ in.). For some types of work, for example, examination of fish in

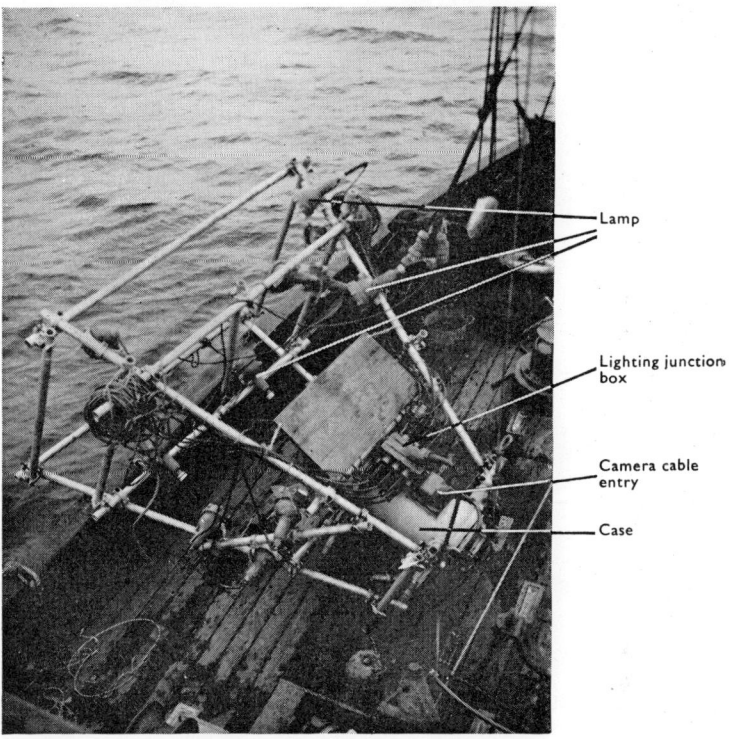

Figure 105

Equipment resting on deck of M.V. *Calanus*

mid-water, the greater part of the framework is dispensed with; the lamps, projecting directly forwards, are mounted round the camera case.

Fourteen standard tungsten Siebe Gorman divers' lamps each with 150-watt floodlight projection bulbs can be mounted anywhere on the gantry; by means of a lighting switchboard mounted adjacent to the main channel controls in the laboratory, the lamps may be switched independently in six pairs and two singles. It has already been stressed (see pp. 173-4) that the correct lighting of the subject is of great importance. A large

Figure 106

Equipment on board M.V. *Calanus*

The camera and lighting cables can be seen leading inboard. The gear is rigged for bottom survey work and is carried lashed to the rail in good weather

frame and control over the individual lamps is therefore a great asset in producing a high picture quality.

The television equipment alone takes about 2 kw. of power at 230 volts, and this is derived from a small convertor, permanently installed in the forward hatch of M.V. *Calanus* (the research ship of the Scottish Marine Biological Association); the lights are run from the ship's D.C. supply.

The assembly and dismantling of the gantry takes a considerable time, particularly in view of the large number of cables joined to the distribution box. Further, when the best arrangement of the lights has been determined for any particular type of work it is desirable to keep the lamps undisturbed. The whole assembly is, therefore, kept together as a unit, being hoisted directly from aboard the ship on to a barrow, on which the apparatus is kept when not in use.

The gear is worked from the fore deck in front of the trawl winch, with all cables flaked in a deep box on the starboard quarter, the inboard ends being taken to the lower laboratory, in which the units are housed and where all the control equipment is installed. It is lowered over the port quarter on a 3·5 cm. non-spin steel rope, worked from a small auxiliary drum fitted to the trawl winch; great care is taken in paying out the cables, particularly the camera cable, because apart from being expensive to renew, any damage to this cable would allow direct ingress of water into the camera case. For convenience in paying out, the camera and lighting cables are kept permanently lashed together at about every 2 metres and, to avoid strain, they are lashed to the suspension wire every 20 metres.

During the descent instructions are always relayed from the lower laboratory to the man at the winch by a microphone and talk-back system, and particular care is taken when nearing the bottom. However, with the ship at anchor danger of hitting bottom and so damaging the gear is small, since the proximity of the bottom is readily appreciated by the increased brightness on the viewing monitor resulting from the reflection of light back into the camera. A further check on the position of the gear can also be made by the echo sounder since an adequate trace of the descending gear is seen.

No attempt has been made to control the orientation of the camera. There are two distinct problems; first, the orientation of anything seen on the viewing monitor may be required with respect to a fixed direction, and secondly, one may wish, particularly when using the camera horizontally, to position it in a selected direction. The second problem involves not only a knowledge of the orientation of the immersed camera but

Figure 107   Inshore ground, Firth of Clyde
Stones, shells and gravel at 10 fathoms. A large anemone (*Bolocera eques*) in the centre. Top right, small crab *Munida bamffica*. Camera about 6 ft. from the ground. (4" lens)

Figure 108   Firth of Clyde
A muddy bottom at 30 fathoms. Note tracks and other surface marking. Fish, bottom right corner. ($2\frac{1}{2}$" lens)

Figure 109　　　　　English Channel
A gravelly-stone bottom (20 fathoms) with dense population of brittle stars (*Ophiothrix fragilis*). Note the variable pattern in the central disc. Camera about 2 ft. from the ground. (4" lens)

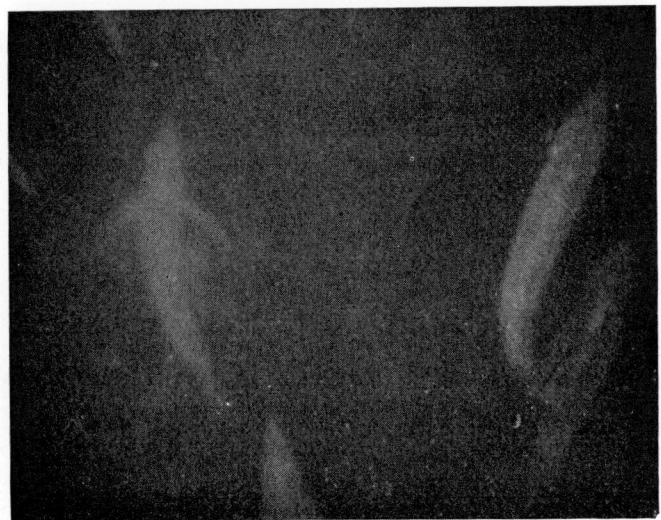

Figure 110　　　　　English Channel
Horse-mackerel (*Trachurus trachurus*). The fish are about 8-10 in. body length and some were 8 ft. away from the vertically rigged camera. (4" lens)

also necessitates a means of rotating it into the required direction. Movement of the camera by remote control about the axis of suspension could readily be obtained by the provision of a remotely controlled underwater motor and fins, as has been done on the Canadian equipment. However, it would seem that in the biological applications of underwater television such control over orientation is not of great importance and no provision for it has been made. To know the orientation of objects seen on the bottom when working vertically is, however, quite important in certain types of marine problems, particularly those of a physical nature. For example, it is of considerable interest to relate the character of the bottom, say orientation of sand ripples or sedentary organisms, to physical topography, to coast-lines or to currents. For such purposes it is only necessary to have a compass mounted in the equipment and to know the direction of north in relation to the particular objects under consideration. Various systems could be devised, some complicated and expensive using remote-indicating compasses, but we have merely mounted an ordinary liquid compass in the field of view of the camera and orientated subjects in relation to magnetic north.

Comparison of television and photography

The problems associated with underwater television are in many ways similar to those of remotely controlled cameras; these problems have already been discussed in detail (see p. 171 *et seq.*). It remains, however, to consider the relative merits of television and photography.

Underwater television has the great advantage over underwater photography that the scene can be viewed continuously and records made at the time by the biologist. The viewing can, so to speak, within the limits of human endurance and the financial and technical resources of maintenance, be virtually continuous. This also has technical advantages; because the scene is continuously under view, adjustments may be made by the remote control mechanisms so that the best picture may be produced of a given scene during viewing—a feature not present in underwater photography. This ensures better picture quality and makes recognition of details in the subject far easier than is the case when they have to be interpreted later from photographs.

The Millport equipment has been used for a variety of investigations—in local waters, the North Sea, the Irish Sea, the English Channel—and representative pictures are shown in Figures 107-110.

EPILOGUE

PROGRESS in any field of natural science is intimately bound up with the technical resources available. Nevertheless, all tools are restricted in their use and as soon as they begin to give diminishing returns it is time to look for new ways of seeking information.

It is a truism that scientific books are never completely up to date when they reach the reading public; this one is no exception. Technical progress in the instrumentation of oceanography continues, and we may briefly note some of the recent developments in the fields that have been covered in the preceding pages.

A new net has been described by Currie and Foxton; this is for vertical hauls and has a depth and water-flow meter in the mouth. The flow meter is a calibrated geared propeller and depth is recorded by a Bourdon spiral as in the bathythermograph. The propeller turns a smoked slide on which the stylus attached to the Bourdon gauge writes, so that a continuous record of water filtered and depth is obtained. In this way any clogging of the net during its ascent becomes known.

Swallow has recently developed a new technique for measuring deep currents; a small simple type of sound transmitter is carried in a long buoy—a cylinder of metal—which can be adjusted to float at a pre-determined depth. This buoy is tracked by means of the sound waves given out by its transducer and picked up by hydrophones hung over the ship's side. J. B. Hersey and his colleagues have described a new echo sounder; in this the transducers can be focused at about 5-10 feet, and when the instrument is under water, echoes can be obtained from single organisms. This instrument has been used to investigate the deep scattering layer and new evidence suggests that fish are responsible for many of the single echoes recorded. Recently this sounder has been mounted on underwater television equipment and, for the first time, it has been possible to 'see' a fish responsible for a specific echo trace.

EPILOGUE

The U.S. Fish and Wildlife Service scientists have recently mounted an underwater television unit both inside and outside a commercial trawl and have been able to watch the behaviour of fish as they pass back into, or escape from, the cod end. New cameras have been described and in one contact with the bottom is indicated by a change of the 'pinging' rate of a small sound source attached to the unit.

REFERENCES

THERE are chapters on equipment in some of the standard textbooks and most expedition reports include an account of the gear used. Except for a Russian text, however, there is no single account of the equipment used in oceanography and marine biology. The following references are, therefore, to journals; they include not only work specifically referred to in the text but other papers of particular interest and importance. For convenience they are collected under the sectional headings of the text.

1. *Sampling the living organisms*

Anderson, A. W., and Lyman, L. (1952). Oceanographic instruments, their use and application in marine biology. *Bull. Gulf and Caribbean Fish. Inst.*, p. 103, April 1952.

Arnold, E. L. (jr.). (1952). High speed plankton samplers. 1. A high speed plankton sampler (Model Gulf 1-A). *U.S. Fish and Wildlife Service, Spec. Sci. Rept., Fish.*, No. 88.

Barnes, H. (1953). A simple and inexpensive closing net. *Mem. Ist. Ital. Idrobiol.*, vol. 7, p. 189.

Boden, B. P. *et alia*. (1955). A depth telerecording unit for marine biology. *J. Mar. Res.*, vol. 14, p. 205.

Clarke, G. L., and Bumpus, D. F. (1940). The plankton sampler—an instrument for quantitative plankton investigations. *Limnol. Soc. of America Spec. Publ.*, No. 5.

Currie, R. I., and Foxton, P. (1956). The Nansen closing method with vertical plankton nets. *J. Mar. biol. Ass. U.K.*, vol. 35, p. 483.

Currie, R. I., and Foxton, P. (1957). A new quantitative plankton net. *J. Mar. biol. Ass. U.K.*, vol. 36, p. 17.

Dakin, W. J. (1907-8). Methods of plankton research. *Trans. Lpool. Biol. Soc.*, vol. 22, p. 500.

Dakin, W. J. (1908). The filtration coefficient of plankton nets. *Lanc. Sea Fish. Lab. Report*, No. 17, p. 126.

Ekman, V. W. (1905). An apparatus for the collection of bottom samples. *Publ. Circ. Cons. Explor. Mer*, No. 27.

Emery, K. O., and Champion, A. R. (1948). Underway bottom sampler. *J. Sediment. Petrol.*, vol. 18, p. 30.

Emery, K. O., and Dietz, R. S. (1941). Gravity coring instrument and

REFERENCES

mechanics of sediment coring. *Bull. geol. Soc. Amer.*, vol. 52, p. 1685.

Forster, G. R. (1953). A new dredge for collecting burrowing animals. *J. Mar. biol. Ass. U.K.*, vol. 32, p. 193.

Gardiner, A. C. (1943). Measurement of phytoplankton populations by the pigment extraction method. *J. Mar. biol. Ass. U.K.*, vol. 25, p. 739.

Gibbons, S. G. (1939). The Hensen net. *J. Cons. int. Explor. Mer*, vol. 14, p. 242.

Gibbons, S. G., and Fraser, J. H. (1937). The centrifugal pump and suction hose as a method of collecting plankton samples. *J. Cons. int. Explor. Mer*, vol. 12, p. 155.

Glover, R. S. (1953). The Hardy Plankton Indicator and sampler: a description of the various models in use. *Hull Bull. Mar. Ecol.*, vol. 4, p. 7.

Hardy, A. C. (1936). The continuous plankton recorder, with an appendix: test of the validity of the continuous plankton recorder method by A. C. Hardy and N. Ennis. *Disc. Rept.*, vol. 11, p. 457.

Hardy, A. C. (1939). Ecological investigations with the continuous Plankton Recorder: object, plan and methods. *Hull Bull. Mar. Ecol.*, vol. 1, p. 1.

Hardy, A. C. (1956). *The Open Sea and the World of Plankton*. Collins, London.

Harvey, H. W. (1934). Measurement of phytoplankton populations. *J. Mar. biol. Ass. U.K.*, vol. 19, p. 761.

Harvey, H. W. (1935). Note concerning a measuring plankton net. *J. Cons. int. Explor. Mer*, vol. 10, p. 179.

Holme, N. A. (1949). A new bottom-sampler. *J. Mar. biol. Ass. U.K.*, vol. 28, p. 323.

Holme, N. A. (1955). An improved 'vacuum' grab for sampling the sea floor. *J. Mar. biol. Ass. U.K.*, vol. 34, p. 545.

Hvorslev, M. J., and Stetson, H. C. (1946). Free-fall coring tube: a new type of gravity bottom sampler. *Bull. geol. Soc. Amer.*, vol. 57, p. 935.

Jenkins, J. T. (1901). The methods and results of the German plankton investigations, with special reference to the Hensen nets. *Trans. Lpool. biol. Soc.*, vol. 15, p. 279.

Kemp, S., Hardy, A. C., and Mackintosh, N. A. (1929). Discovery investigations. Objects, equipment and methods. *Disc. Rept.*, vol. 1, p. 141.

Kofoid, C. A. (1911). On a self-closing plankton net for horizontal towing. On an improved form of a self-closing water bucket for plankton investigation. *Univ. Calif. Publ. Zool.*, vol. 8, p. 311.

Kullenberg, B. (1947). The piston core sampler. *Svenska hydrogr.-biol. Komm. Skr., Ser. Hydrogr.*, vol. 1, p. 2.

Kullenberg, B. (1955). A new core-sampler. *Göteborgs Kungl. vetenskaps-och vitterhets samhälles Handl., Sjatte Följden, Ser.*, vol. 6, No. 15, 17 pp.

McIntyre, A. D. (1956). The use of trawl, grab and camera in estimating marine benthos. *J. Mar. biol. Ass. U.K.*, vol. 35, p. 419.

Moore, H. B., and Neill, R. G. (1930). An instrument for sampling marine muds. *J. Mar. biol. Ass. U.K.*, vol. 16, p. 589.

Nielsen, E. S. (1933). Uber quantitative Untersuchung von marinen Plankton mit Utermohls umgekehrtem Mikroskop. *J. Cons. int. Explor. Mer*, vol. 8, p. 201.

Ostenfeld, C. H., and Jespersen, P. (1924). Standard net for plankton collections. *Publ. Circ. Cons. Explor. Mer*, No. 84.

Petersen, C. G. J. (1918). The sea bottom and its production of fish food. I. Apparatus for investigation of the sea bottom. *Danish biol. Station Rept.*, No. 25, p. 1.

Pettersson, O. (1928). A new apparatus for the taking of bottom samples. *Sven. Hydrogr.-Biolog. Komm. Skr., N.S. Hydrogr.* Vol. 4, p. 6.

Piggot, C. S. (1937). Core samples of the ocean bottom. *Ann. Rept. Smithsonian Inst.*, 1936, p. 207.

Pratje, O. (1950). Eine neue Lotröhre und ihre erste Erprobung. *Dtsch. hydrogr. Z.*, vol. 3, p. 100.

Smith, O. R., and Ahlstrom, E. H. (1948). Echo-ranging for fish schools and observations on temperature and plankton in waters off central California in the Spring of 1946. *U.S. Fish and Wildlife Service, Spec. Sci. Rept., Fish.*, No. 44.

Smith, W., and McIntyre, A. D. (1954). A spring-loaded bottom sampler. *J. Mar. biol. Ass. U.K.*, vol. 33, p. 257.

Tonolli, V. (1951). Un nuovo apparecchio per la cattura di plancton in modo continuo e quantitativo : il 'Plancton-Bar'. *Mem. Ist. Ital. Idrobiol.*, vol. 6, p. 193.

Wiborg, K. F. (1951). The whirling vessel. An apparatus for the fractionating of plankton samples. *Rep. Norweg. Fish. and Mar. Investig.*, vol. 9, No. 13.

Wickstead, J. (1953). A new apparatus for collecting bottom plankton. *J. Mar. biol. Ass. U.K.*, vol. 32, p. 347.

REFERENCES

Wimpenny, R. S. (1928). A transparent bucket with a detachable bottom for use with Ostenfeld and Jespersen's standard net for plankton collection. *J. Cons. int. Explor. Mer*, vol. 3, p. 94.

Wimpenny, R. S. (1937). A new form of Hensen net bucket. *J. Cons. int. Explor. Mer*, vol. 12, p. 178.

ZoBell, C. E. (1941). Apparatus for collecting water samples from different depths for bacteriological analysis. *J. Mar. Res.*, vol. 4, p. 173.

2. The use of sound waves

Balls, R. (1934). A revolutionary development in herring fishing. *Fish Trades Gazette*, Nov. 24th, p. 19.

Balls, R. (1946). Fish on the spotline. *Marconi Int. Mar. Comm. Co. Ltd.* London, Leaflet.

Balls, R. (1951). Environmental changes in herring behaviour: a theory of light avoidance, as suggested by echo-sounding observations in the North Sea. *J. Cons. int. Explor. Mer*, vol. 17, p. 274.

Craig, R. E. (1954). Echo-sounding in marine biology. *Advanc. Sci. Lond.*, vol. 11, p. 51.

Craig, R. E. (1945). The use of echo-sounder in fish location—a survey of present knowledge, with notes on the use of Asdic. *Rapp. Proc-Verb.*, vol. 139, App. II, p. 32.

Cushing, D. H., Devold, F., Marr, J. C., and Kristjonsson, H. (1952). Some modern methods of fish detection. Echo-sounding, echo-ranging and aerial scouting. *F. A. O. Fish. Bull.*, vol. 5, No. 3-4, 27 pp.

Cushing, D. H., and Richardson, I. D. (1955). Echo-sounding experiments on fish. *Fish. Investig. Lond.*, Ser. II, vol. 18, No. 4, 34 pp.

Dietz, R. S. (1948). Deep scattering layer in the Pacific and Antarctic Oceans. *J. Mar. Res.*, vol. 7, p. 430.

Everest, F. A., Young, R. W., and Johnson, M. W. (1948). Acoustical characteristics of noise produced by snapping shrimp. *J. acoust. Soc. Amer.*, vol. 20, p. 137.

Fish, M. P. (1954). The character and significance of sound production among fishes of the western North Atlantic. *Bull. Bingham Oceanogr. Coll.*, vol. 14, Art. 3, 109 pp.

Fisher, A. (1821). In W. E. Parry. *Journal of a Voyage for the discovery of a north-west passage from the Atlantic to the Pacific ... H.M.S. Hecla and Griper.* London, 1821, p. 35.

Fisher, A. (1821). *A journal of a voyage of discovery to the Arctic regions in H.M.S. 'Hecla' and 'Griper' in the years 1819 and 1820.* London, 1821, p. 73.

Hersey, J. B., and Moore, H. B. (1948). Progress report on scattering layer observations in the Atlantic Ocean. *Trans. Amer. Geophys. Union*, vol. 29, p. 341.

Hodgson, W. C. (1950), Echo-sounding and the pelagic fisheries. *Fish. Investig. Lond.*, Ser. II, vol. 17, No. 4.

Johnson, M. W. (1948). Sound as a tool in marine ecology from data on biological noises and the deep scattering layer. *J. Mar. Res.*, vol. 7, p. 443.

Johnson, M. W., Everest, F. A., and Young, R. W. (1947). The role of snapping shrimp (*Crangon* and *Synalpheus*) in the production of underwater noise in the sea. *Biol. Bull.*, Woods Hole, vol. 93, p. 122.

Kane, E. K. (1854). *The U.S. Grinnell Expedition in search of Sir John Franklin.* New York, 1854, p. 359.

Marshall, N. B. (1951). Bathypelagic fishes as sonic scatterers. *J. Mar. Res.*, vol. 10, p. 1.

Moore, H. B. (1950). The relation between the scattering layer and the Euphausiacea. *Biol. Bull.*, Woods Hole, vol. 99, p. 181.

Raitt, R. W. (1948). Sound scatterers in the sea. *J. Mar. Res.*, vol. 7, p. 393.

Richardson, I. D. (1952). Some reactions of pelagic fish to light as recorded by echo-sounding. *Fish. Investig. Lond.*, Ser. II, vol. 18, No. 1.

Schevill, W. E., and Lawrence, B. (1949). Underwater listening to the white porpoise (*Delphinapterus leucas*). *Science*, vol. 109, p. 143.

Tucker, G. H. (1951). Relation of fishes and other organisms to the scattering of underwater sound. *J. Mar. Res.*, vol. 10, p. 215.

3. *Some properties of the water itself*

Anderson, E. R., and Burke, A. T. (1951). Notes on the development of a thermistor temperature profile recorder (TPR). *J. Mar. Res.*, vol. 10, p. 168.

Arx, W. S. von (1950). An electromagnetic method for measuring the velocities of ocean currents from a ship underway. *Mass. Inst. Technol. Meteorol. Papers*, vol. 11, p. 3.

Arx. W. S. von. (1950). Some surface measurements of ocean current velocities obtained from a ship underway by means of the geo-

REFERENCES

magnetic electro-kinetograph. *Trans. Amer. Geophys. Union,* vol. 31, p. 331.

Arx, W. S. von (1950). Some current meters designed for suspension from an anchored ship. *J. Mar. Res.,* vol. 9, p. 93.

Bidder, G. P. (1905). Account of some experiments on bottom trailers. *Rapp. Proc.-Verb.,* vol. 4, July.

Bowden, K. F. (1954). The direct measurement of subsurface currents in the oceans. *Deep-Sea Res.,* vol. 2, p. 33.

Brooks, C. F. (1926). Observations on sea temperatures. U.S. Dept. of Agric., *Weather Bur. Monthly Weather Rev.,* vol. 54, p. 241.

Brooks, C. F. (1928). Reliability of different methods of taking sea surface temperatures. *J. Wash. Acad. Sci.,* vol. 18, p. 525.

Carruthers, J. N. (1924). A new drift indicator. *Nature. Lond.,* vol. 114, p. 718.

Carruthers, J. N. (with Garrood, H. J., and Edser, J.) (1926). A new current measuring instrument for the purposes of fishery research. *J. Cons. int. Explor. Mer,* vol. 1, p. 127.

Carruthers, J. N. (1928). New drift bottles for the investigation of currents in connection with fishery research. *J. Cons. int. Explor. Mer,* vol. 3, p. 194.

Carruthers, J. N. (1928). The flow of water through the Straits of Dover as gauged by continuous current meter observations at the Varne Lightvessel (50° 56' N. 1° 17' E.). *Fish. Investig. Lond.,* Ser. II. vol. 11, No. 1, 109 pp.

Carruthers, J. N. (1930). Further investigations upon the water movements in the English Channel. Drift bottle experiments in the summers of 1927, 1928 and 1929, with critical notes on drift bottle experiments in general. *J. Mar. biol. Ass. U.K.,* vol. 17, p. 241.

Carruthers, J. N. (1935). A 'vertical log' current meter. *J. Cons. int. Explor. Mer,* vol. 10, p. 151.

Carruthers, J. N. (1947). Practical proposals for a continuous programme of thick-layer current measuring in all weathers, with remarks on relevant wind observations and other related matters. *J. Cons. int. Explor. Mer,* vol. 15, p. 13.

Carruthers, J. N. (1955). Some simple oceanographical instruments to aid in certain forms of commercial fishing and in various problems of fisheries research. F.A.O. *Fisheries Bull.,* vol. 8, No. 3, p. 130.

Church, P. E. (1932). Surface temperatures of the Gulf Stream and its bordering waters. *Geographical Rev.,* vol. 22, p. 286.

Ekman, V. W. (1905). On the use of insulated water bottles and reversing thermometers. *Publ. Circ. Cons. Explor. Mer.*, No. 23.
Ekman, V. W. (1926). On a new repeating current meter. *Publ. Circ. Cons. Explor. Mer*, No. 91.
Ekman, V. W. (1932). An improved type of current meter. *J. Cons. int. Explor. Mer*, vol. 7, p. 1.
Garbell, M. A. (1947). Fins for aerological instruments. *J. Meteorol.*, vol. 4, p. 82.
Hamon, B. V. (1955). A temperature-salinity depth recorder. *J. Cons. int. Explor. Mer*, vol. 21, p. 72.
Kalle, K. (1948). A new method for measuring surface currents at sea from anchored ships. *Dtsch. hydrogr. Z.*, vol. 1, p. 164.
Kirk, T. H., and Gordon, A. H. (1952). Comparison of intake and bucket method for measuring sea temperature. *Mar. Observer*, vol. 22, p. 33.
Knudsen, M. (1929). A frameless reversing water bottle. *J. Cons. int. Explor. Mer*, vol. 4, p. 192.
Lumby, J. R. (1927). The surface sampler, an apparatus for the collection of samples from the sea surface from ships in motion. *J. Cons. int. Explor. Mer*, vol. 2, p. 332.
Lumby, J. R. (1928). Modification of the surface sampler with a view to the improvement of temperature observation. *J. Cons. int. Explor. Mer*, vol. 3, p. 340.
Lumby, J. R. (1929). A surface sampler for temperature observations. *J. Cons. int. Explor. Mer*, vol. 4, p. 281.
Ministry of Agriculture and Fisheries (1932). Current meter work. *Fisheries Notice*, No. 17.
Ministry of Agriculture and Fisheries (1931). Drift bottles. *Fisheries Notice*, No. 16.
Ministry of Agriculture and Fisheries (1939). Current meter work. II. The vertical log current meter and its use. *Fisheries Notice*, No. 26.
Mosby, H. (1944). The thermo sound. An oceanographic temperature recorder. Bergens Museum Årbok (1943). *Naturv. rekke*, No. 1, p. 1.
Olson, F. C. W. (1951). A plastic envelope substitute for drift bottles. *J. Mar. Res.*, vol. 10, p. 190.
Olson, R. A. (1941). A rapid response thermocouple of high sensitivity for the determination of temperature stratification in natural waters. *Chesapeake Bay Biol. Lab., Publ.* No. 45.

REFERENCES

Pettersson, O. (1905). Beschreibung des Bifilar-Strommessers. *Publ. Circ. Cons. Explor. Mer*, No. 25.

Pettersson, O. (1929). Current meter for determination of the direction and velocity of the movement of water at the bottom of the ocean. *Sven. Hydrogr.-Biolog. Komm. Skr. N.S. Hydrogr.*, vol. 13, p. 9.

Pritchard, D. W., and Burt, W. V. (1951). An inexpensive and rapid technique for obtaining current profiles in estuarine waters. *J. Mar. Res.*, vol. 10, p. 180.

Spilhaus, A. F. (1937-8). A bathythermograph. *J. Mar. Res.*, vol. 1, p. 95.

Spilhaus, A. F. (1940). A detailed study of the surface layers of the ocean in the neighbourhood of the Gulf Stream with the aid of rapid measuring hydrographic instruments. *J. Mar. Res.*, vol. 3, p. 51.

Spilhaus, A. F. (1949). Bathythermograph sea sampler. *Ass. d'Océanogr. Phys. Union Géodés. et Géophys. Int., Proc.-Verb.*, No. 4, p. 146.

Spilhaus, A. F., and Miller, A. R. (1948). The sea sampler. *J. Mar. Res.*, vol. 7, p. 370.

Strøm, K. M. (1939). A reversing thermometer by Richter and Wiese with $1/100\text{th}°$ C. gradation. *Int. Rev. Hydrobiol.*, vol. 38, p. 259.

Tait, J. B. (1934). Surface drift bottle results in relation to temperature, salinity and density distributions in the Northern North Sea. *Rapp. Proc-Verb.*, vol. 89, p. 69.

Tait, J. B. (1932). The surface water drift in the northern and middle areas of the North Sea and in the Faroe-Shetland Channel. Part II. Section 2. A cartographical analysis of the results of Scottish surface drift bottle experiments commenced in the year 1911. Fishery Bd. Scotland, *Sci. Investig.* 1931, No. 3.

Tully, J. P. (1937). A graphical method for calculating the corrections on deep-sea reversing thermometers. *J. Cons. int. Explor. Mer*, vol. 12, p. 40.

Vaux, D. (1955). Current measuring in shallow waters by towed electrodes. *J. Mar. Res.*, vol. 14, p. 187.

Witting, R. (1932). A current meter, its use and some results. *J. Cons. int. Explor. Mer*, vol. 7, p. 218.

4. *Photography and Television*

Anon. (1952). Underwater photography of the seine net whilst fishing. *World Fishing*, vol. 1, p. 329.

Backus, R. H., and Barnes, H. (1957). Television-echo sounder observations of midwater sound scatterers. *Deep-Sea Res.*, vol 4, p. 116.

Baird, R. H., and Gibson, F. A. (1956). Underwater observations on escallop (*Pecten maximus* L.) beds. *J. Mar. biol. Ass. U.K.*, vol. 35, p. 555.

Baker, A. de C. (1957). Underwater photographs in the study of oceanic squid. *Deep-Sea Res.*, vol. 4, p. 126.

Barnes, H. (1952). Underwater television and marine biology. *Nature. Lond.*, vol. 169, p. 477.

Barnes, H. (1952). Television for marine research. *The Listener*, Dec. 25th, 1952.

Barnes, H. (1953). Underwater television and the fisheries. *The Fishing News*, No. 2089, May 2nd, 1953.

Barnes, H. (1953). Underwater television and marine research. *Discovery*, vol. 14, June 1953.

Barnes, H. (1953). Underwater television and research in marine biology, bottom topography and geology. Part I. A description of the equipment and its use on board ship. *Dtsch. hydrogr. Z.*, vol. 6, p. 123.

Barnes, H. (1954). Symposium on New Advances in Underwater Observations. *Advanc. Sci. Lond.*, vol. 11, No. 41, p. 49.

Barnes, H. (1954). Underwater television pictures. *Discovery*, vol. 15, No. 5.

Barnes, H. (1955). Underwater television and research on marine ecology, bottom topography and geology. Part 2. Experience with the equipment. *Dtsch. hydrogr. Z.*, vol. 8, p. 213.

Barnes, H. (1956). Underwater television and dockyard practice. *Journ. Inst. Civil Eng.*, Nov. 1956.

Chesterman, W. D. (1950). Photography under the sea. *Funct. Photogr.*, vol. 1, p. 5.

Collins, J. B. (1950). Underwater photography. *Photogr. J.*, vol. 90B, p. 24.

Emery, K. O. (1952). Submarine photography with the Benthograph. *Sci. Mon., N.Y.*, vol. 75, p. 3.

Ewing, M., Vine, A. C., and Worzel, J. L. (1946). Photography of the ocean bottom. *J. opt. Soc. Amer.*, vol. 36, p. 307.

Hahn, J. (1950). Some aspects of deep-sea underwater photography. *J. Phot. Soc., Amer. Sect. B. Photo. Sci. Tech.*, vol. 16, p. 27.

Harvey, E. N., and Baylor, E. R. (1948). Deep-sea photography. *J. Mar. Res.*, vol. 7, p. 10.

REFERENCES

Laughton, A. S. (1957). A new deep-sea underwater camera. *Deep-Sea Res.*, vol. 4, p. 120.

Margetts, A. R. (1952). Some conclusions from underwater observation of trawl behaviour. *World Fishing*, vol. 1, p. 161.

Swallow, J. C. (1957). Some further deep current measurements using neutrally-buoyant floats. *Deep-Sea Res.*, vol. 4, p. 93.

INDEX

Aerial photography, 159-167
Aerial photography and fisheries, 167-169
Aerial scouting, 166-167, 169
Agassiz trawl, 47-48
Ahlstrom, E. H., 206
Allan Hancock Foundation, 180
Amplification, 78-79
Anchor dredge, 49
Anderson, A. W., 204
Anderson, E. R., 208
Antarctic Ocean, 20, 34, 83, 86
Aqualung, 167
Arnold, E. L. (Jr.), 204
Arrago, Jean F., 72
Asdic, 94
Atlantic Ocean, 43-44, 71, 110-111, 132

Backus, R. H., 212
Bacteria, 16-19; counting of, 18
Baird, R. H., 212
Baker, A. de C., 212
Balls, R., 87, 88, 207
Barnes, H., 204, 212
Bathythermograph, 121-127
Baylor, E. R., 212
Beach profiles, 159-166
Beam trawl, 47-48
Benthograph, 180-183
Bidder, G. P., 209
Boden, B. P., 204
Bokn, Skipper, 87
Bottom fauna, 46, 171-173
Bottom sediments, 46, 170-171
Bottom trailing bottles, 131-132
Bourdon spiral, 121-127
Bowden, K. F., 209
Brightness profile, 160-166
British Army Photographic Research Unit, 160

Brooks, C. F., 107, 108, 209
Bumpus, D. F., 204
Burke, A. T., 208
Burt, W. V., 211

Calanus finmarchicus, 41
Canadian underwater television, 200
Candacia armata, 42-43
Carruthers, J. N., 145, 148, 149, 150, 151, 152, 209
Carruthers' drift indicator, 148-150; vertical log, 148, 151-152
Cathode ray tube, echo sounders, 80, 82-83; scanning, 187-188
Champion, A. R., 59, 204
Chesapeake Bay, 96-98
Chesterman, W. D., 212
Chrysochromalina minor, 19
Church, P. E., 110, 111, 209
Clarke, B., 19
Clarke, G. L., 204
Clarke-Bumpus sampler, 26-29
Clogging of nets, 22-23, 40-41
Closing devices, 26
Closing net, 26
Coalfish, 89
Cod, 87, 89
Collins, J. B., 212
Conductivity cells, 121
Continuous plankton recorder, 34-46
Continuous recording current meters, 139-158
Core, 71
Core-length, 67
Corer, Emery-Dietz gravity, 62-64; Hvorslev and Stetson, 64, 66; Kullenberg piston, 67-71; Moore and Neill, 57-59

214

INDEX

Corethron criophilum, 20
Cornell Aeronautical Laboratory, 183
Counting, bacteria, 18, 19; diatoms, 21; zooplankton, 32, 33
C.P.S. Emitron tube, 186, 189-190
Craig, R. E., 176, 177, 207
Crangon, 99-100
Croaker noise, 96-97
Current meter, Ekman, 134-139; Jacobsen, 145-148; von Arx, 140-144
Current meters, continuous recording, 139-152
Currents, methods of measurement, 127-158, 202; continuous recording of, 139-144; flow methods, 134-139; long period measuring, 144-152
Currie, R. I., 202, 204
Cushing, D. H., 90, 207

Dakin, W. J., 204
Decapod larvae, 43-45
Decca, 134
Dechevrens, Marc, 153
Deep scattering layer, 93-94, 202
Deep-sea cameras, 178-183
Devold, F., 207
Diatoms, 16, 20-22; counting of, 21
Dietz, R. S., 62, 63, 64, 204, 207
'Discovery' Expeditions, 34, 83, 86
Dogger Bank, 44, 45
Dredge, anchor, 49
Dredges, 49
Drift, 128
Drift bottles, 129-134
Drift indicator, Carruthers', 148-150

Earth's magnetic field, 152
Echo signals, recording of, 79-83
Echo sounder, 72-94, 202; and fisheries, 83-87

Echo trace and fish species, 88-91
Edser, J., 209
Eicher, G. J., 168, 169
Ekman, F. L., 62
Ekman, V. W., 62, 204, 210
Ekman current meter, 134-139
Electrokinetograph, geomagnetic, 153-158
Electromagnetic induction, 140, 152
Emery, K. O., 59, 61, 62, 63, 64, 65, 169, 170, 182, 204, 212
Emery-Dietz gravity corer, 62-64
E.M.I. Research Laboratories Ltd., Hayes, 189
English Channel, 132-133
Envelopes, plastic, 133
Everest, F. A., 103, 104, 207, 208
Ewing, M., 178, 212

Faraday, M., 152, 153
Filtration coefficient, 22-23
Fish, M. P., 207
Fish noise, 96-98, 100-102
Fisher, A., 94, 95, 207, 208
Forster, G. R., 49, 205
Foxton, P., 202, 204
Fraser, J. H., 205
Frogmen, 168

Garbell, M. A., 210
Gardiner, A. C., 205
Garrood, H. J., 209
Geomagnetic Electrokinetograph (G.E.K.), 153-158
Gibbons, S. G., 205
Gibson, F. A., 212
Glover, R. S., 31, 205
Gordon, A. H., 210
Grab, Holme, 51-52; Petersen, 51; Smith-McIntyre, 53-55; vacuum, 54, 56-57; van Veen, 51
Grabs, 50-54
Gravity-corer, Emery-Dietz, 62-65
Gulf Stream, 110, 127, 156-158

215

Haemocytometer, 18, 21
Hahn, J., 212
Hamon, B. V., 210
Hardy, Sir Alister C., 33, 34, 36, 205
Hardy plankton indicator, 28-31
Harvey, E. N., 212
Harvey, H. W., 21, 205
Hauls, horizontal, 23-26; vertical, 23
Heezen, B. C., 171, 172
Hendey, N. Ingram, 20
Hensen, V., 22, 32
Hensen net, 24
Herdman, H. F. P., 86
Herring, 87-89, 91-92
Hersey, J. B., 202, 208
Hodgson, W. C., 89, 208
Holme, N. A., 51, 52, 56, 205
Holme sampler, 51-52
Horizontal hauls, 23-26
Hvorslev, M. J., 66, 71, 205
Hvorslev and Stetson corer, 66
Hydrographic Service, British Navy, 72, 73
Hydrophone techniques, 73, 95, 102-105

Image orthicon tube, 188, 189
Insulated water-bottle, 110-113
Inverted microscope, 21

Jacobsen current meter, 145-148
Jenkins, J. T., 205
Jespersen, P., 206
Johnson, M. W., 103, 104, 207, 208

Kalle, K., 210
Kane, E. K., 95, 208
Kelez, G. B., 167
Kelvin and Hughes, 75-79, 84-85
Kemp, S., 205
Kirk, T. H., 210
Knudsen, M., 112, 210

Knudsen frameless water-bottle, 116-119
Kofoid, C. A., 26, 206
Kristjonsson, H., 207
Kullenberg, B., 62, 68, 69, 70, 206
Kullenberg piston-corer, 67-71

Langevin, P., 74
Laughton, A. S., 213
Lawrence, B., 208
Light penetration, 164
Lighting for underwater cameras, 173-175
Loran, 134
Loye, D. P., 97, 98
Lumby, J. R., 108, 109, 210
Lumby sampler, 108-110
Lyman, L., 204

McIntyre, A. D., 53, 54, 55, 206
Mackerel, 89
Mackintosh, N. A., 205
Magneto-striction effect, 74-75, 77
Manton, I., 19
Marconi Company, 81, 87, 184
Margetts, A. R., 213
Marr, J. C., 207
Marshall, N. B., 208
Messenger, 18, 26, 28, 113, 114, 117, 118
Microscope, inverted, 21
Miller, A. R., 211
Millport Marine Station, 183
Ministry of Agriculture, Fisheries and Food, Lowestoft, 24, 25, 88, 89, 92, 137, 146, 210
Modern echo sounder, 75-83
Moore, H. B., 57, 58, 93, 206, 208
Moore, J. Grange, 160, 164
Moore and Neill Corer, 57-59
Mosaic, 186, 187
Mosby, H., 119, 210
Mosby thermo-sound, 119-121
μ-flagellates, 19-20
Munk, W. H., 161

INDEX

Nannoplankton, 16, 19-20
National Institute of Oceanography, 86, 145
Neill, R. G., 57, 58, 206
Nets, Hensen, 24; plankton, 22-32; stramin, 25
Nielsen, E. S., 206
Noise, croaker, 96-97; fish, 100-102; shrimp, 98-100; whale, 94-95
North Sea, 42-44, 132-133

Olson, F. C. W., 210
Olson, R. A., 210
Ostenfeld, C. H., 206
Otter trawl, 47

Paravane, 26
Parke, M., 19
Penetration, coring, 66-67
Penetration of light, 164
Peridinians, 20
Petersen, C. G. J., 50, 206
Pettersson, O., 139, 206, 211
Phleger, F. B., 71
Photography, aerial, 159-167; underwater, 167-183
Photography and underwater television, 200
Piezo-electric effect, 73-74
Piggot, C. S., 62, 206
Pigment extracts, 21
Pigment unit, 21-22
Pilchard, 89
Piston-corer, Kullenberg, 67-71
Plankton, 15-46; nets, 22-32; pumps, 31
Plastic envelopes, 133
Plating method, 18, 19
Plymouth Laboratory of the Marine Biological Association, U.K., 49
Pollack, 89
Pratje, O., 67, 206
Principles of television camera, 184-188
Pritchard, D. W., 211

Production of ultrasonic sound waves, 73-75
Protected thermometers, 116
Protozoa, 20
Proudfoot, D. A., 97, 98

Radio buoys, 134
Rae, K. M., 30, 34, 37, 38, 42, 43
Raitt, R. W., 208
Receiving oscillator, 78
Recording of echo signals, 79-83
Rees, C. B., 42, 43, 45
Resistance thermometers, 121
Reversing thermometers, 115-116
Reversing water-bottle, 113-119
Richardson, I. D., 91, 92, 207, 208
Royal Naval Scientific Service, 184
Runnstrøm, S., 91

Salinity, 106, 107, 119, 121, 125, 127
Sampler, Clarke-Bumpus, 26-29; Lumby, 108-110; underway, 59-62
Scanning, 187-188
Schevill, W. E., 208
Scientific Service, Royal Naval, 184
Scope of underwater cameras, 168-172
Scottish Home Department, Aberdeen, 88, 168
Scottish Marine Biological Association, 35, 184
Scripps Institution, La Jolla, California, 179
Sea sampler, 121-127
Seine net, 168
Shallow-water underwater camera, 176-178
Shrimp noise, 98-100
Smith, O. R., 206
Smith, W., 53, 54, 55, 206
Smith-McIntyre sampler, 53-55
Snapping shrimp, 98-105
Speed of sound in water, 72
Spilhaus, A. F., 121, 211

217

Sponge beds, 102
Sprat, 89
Stempel pipette, 32
Step wedge, 163-164, 173
Stetson, H. C., 66, 71, 205
Stramin net, 25
Strøm, K. M., 211
Sub-surface temperatures, 110-127
Sund, Oscar, 87
Surface temperatures, 107-110
Swallow, J. C., 202, 213
Synalpheus, 99-100
Synoptic plankton picture, 33-46

Tait, J. B., 211
Taylor, M. A., 161
Television, underwater, 183-201
Temperature, sub-surface, 110-127; surface, 107-110
Tester, A. L., 91
Thermistors, 121
Thermometers, protected, 116; resistance, 121; reversing, 115-116
Tonolli, V., 206
Transmitting oscillator, 75-78
Trawl, Agassiz, 47; beam, 47, 48; otter, 47
Trawls and dredges, 47-50, 203
Tucker, G. H., 208
Tully, J. P., 211

Ultrasonic sound waves, 73-75
Underwater cameras for deep water, 178-183; shallow water, 176-178; scope, 169-172
Underwater noise, 94-105
Underwater photography, 167-183
Underwater television, 183-201, 202; and underwater photography, 200; camera and controls, 189-194; casing, 193-194; lighting unit, 195-196; principles of, 184-187; Canadian, 200; Pictures, 198-199
Underway sampler, 59-62
U.S. Navy Electronics Laboratory, 179
USNEL deep-sea camera, 178-181
University College, Hull, 35

Vacuum sampler, 54, 56-57
van Veen, J., 51
van Veen grab, 51
Vaux, D., 211
Vertical hauls, 23, 24
Vertical log, Carruthers', 148-152
Vertical migration, 41-42
Vine, A. C., 178, 212
von Arx, W. S., 140, 141, 142, 143, 153, 154, 155, 156, 158, 208, 209
von Arx current meter, 140-144

Water-bottle, for bacterial sampling, 16-18; insulated, 110-114; Knudsen frameless, 116-119; reversing, 113-119
Whale noise, 94-95
Whales, 20, 94-95
Whirling vessel, 33
Wiborg, K. F., 206
Wickstead, J., 206
Wimpenny, R. S., 207
Witting, R., 139, 211
Wood, A. B., 74
Woods Hole, 153
Worzel, J. L., 178, 212

Young, R. W., 103, 104, 207, 208

ZoBell, C. E., 16, 17, 207
Zooplankton, 22-46; counting, 32

GEORGE ALLEN & UNWIN LTD
London: 40 Museum Street, W.C.1

Auckland: P.O. Box 36013, Northcote Central, N.4
Barbados: P.O. Box 222, Bridgetown
Beirut: Deeb Building, Jeane d'Arc Street
Bombay: 15 Graham Road, Ballard Estate, Bombay 1
Buenos Aires: Escritorio 454-459, Florida 165
Calcutta: 17 Chittaranjan Avenue, Calcutta 13
Cape Town: 68 Shortmarket Street
Hong Kong: 105 Wing On Mansion, 26 Hancow Road, Kowloon
Ibadan: P.O. Box 62
Karachi: Karachi Chambers, McLeod Road
Madras: Mohan Mansions, 38c Mount Road, Madras 6
Mexico: Villalongin 32, Mexico 5, D.F.
Nairobi: P.O. Box 30583
New Delhi: 13-14 Asaf Ali Road, New Delhi 1
Ontario: 81 Curlew Drive, Don Mills
Philippines: Manila, P.O. Box 4322
Rio de Janeiro: Caixa Postal 2637-Zc-00
Singapore: 36c Prinsep Street, Singapore 7
Sydney, N.S.W.: Bradbury House, 55 York Street
Tokyo: P.O. Box 26, Kamata

HAROLD BARNES, *Editor*

OCEANOGRAPHY AND MARINE BIOLOGY

AN ANNUAL REVIEW   VOLUMES 1-6

This series has the following objects—to consider annually basic aspects of marine research returning to each in future volumes at appropriate intervals, to deal with subjects of special and topical importance, and to add new ones as they arise. The favourable reception accorded to Volumes 1 to 6 of the series, which is now separately published in the U.S.A., shows that it is fulfilling a very real need; both reviews and sales have been gratifying. Each volume follows closely the object and style of the first volume continuing to regard the marine sciences with all their various aspects as a unity. Each article endeavours to cover completely the literature of its subject. Physical, chemical and biological aspects of marine science are all dealt with by experts actively engaged in their own field. The series has become an essential reference text for research workers and students and finds a place not only in the libraries of marine stations and fisheries institutes, but also in universities.

Author, Systematic and Subject Indexes

SOME CONTEMPORARY STUDIES IN MARINE SCIENCE

A COLLECTION OF ORIGINAL SCIENTIFIC PAPERS PRESENTED TO DR. S. M. MARSHALL, F.R.S., IN RECOGNITION OF HER CONTRIBUTION WITH THE LATE DR. A. P. ORR TO MARINE BIOLOGICAL PROGRESS.

The Volume contains fifty *original* papers and has become an essential part of the libraries of Marine and Fisheries Laboratories, of Universities, and of all Institutes of higher education.

GEORGE ALLEN & UNWIN LTD